COLLINS GEM GUIDES

FRUITS, NUTS AND BERRIES
AND
CONSPICUOUS SEEDS

Marjorie Blamey
Philip Blamey

COLLINS
London and Glasgow

First published 1984

© Philip and Marjorie Blamey

ISBN 0 00 458822 3

Colour reproduction by Bridge Graphics Ltd, Hull

Filmset by Peter MacDonald Typesetting, Twickenham

Printed and bound by Wm Collins Sons and Co Ltd, Glasgow

Reprint 10 9 8 7 6 5 4 3

Contents

Introduction and about this book

The object of this book is to extend the plant-finding 'season' beyond the flowering time and to add another purpose to country walks in the autumn.

Many plants retain their fruits far longer than the flowers which precede them. This book will help you to identify them after the flowers have gone. Never take the flower to the book but ensure that you always have this Gem and the Wild Flower Gem in your pocket.

The basic parts of the flower are the **anther**, which is the pollen-bearing tip of the **stamen**, and the **stigma** which is the receptive part of the **style** leading to the **ovary** where the seeds develop. **Sepals** grow immediately behind the flower and make up the **calyx** just as the **petals** make the **corolla**. Leaflike structures behind the flower or at the base of the flower stalk are called **bracts**.

Pollination is achieved in many ways, often by insects seeking nectar; some plants produce clouds of pollen and rely on wind dispersal. After pollination the flower fades and the fruit begins to swell. When ripe the seeds such as Thistledown may be blown by the wind or, like Burdock, become caught in the fur of a passing animal. Some, like dandelions, have feathery attachments or, like maples, have wings. Other plants fling their seeds over quite long distances – gardeners who have Bittercress as a weed will know this tiresome habit. Birds disperse seeds in pellets or, like foxes and mice, leave them in their droppings. Some seeds float and are cast ashore by wind or the flow of water on to a muddy bank where they will germinate and grow; alders do this, for example. Erodiums have a 'corkscrew' on their seed which varies its twist with changes in humidity, thus effectively screwing it into the soil.

Having identified a plant it is wise to know more about its fruit. Some plants are deadly poisonous, some less so and some are beneficial: details of these are in the main text. Poisonous plants are marked with a Ⓟ. After handling these plants you would be wise to wash before touching food, and *never* let children eat berries or seeds unless you are certain of their identity.

Some species are rare and are protected under the 1981 Wildlife and Countryside Act which makes it illegal to *dig up any* wild plant without the specific permission of the owner and illegal to remove *any part* of those plants which are on a 'specially protected' list. Fines of up to £500 *per plant* may be imposed. Other species are being added to this list as habitats are destroyed. Apply to your County Trust for Nature Conservation for an up to date list.

Some important families provide man with valuable plants for many purposes. Roses include strawberries, blackberries, apples, plums, cherries and hips for the vitamin C rich syrup. The Carrot family gives us roots like parsnip, and spices like caraway and coriander also come from the Carrot family. The cabbages provide sprouts, cauliflowers and beets, and marjoram and thyme come from the Mint family. Some families like the Nightshades have a wide range of uses. Within this versatile group are the deadly poisonous alkaloid solanine from the berries of Bittersweet; the drug atropine, used by the ophthalmic surgeon, comes from nightshade, and for the cook there are potatoes and tomatoes. The Tobacco plant provides nicotine for those who like it.

Plants throughout Europe are being destroyed by modern progress and the need to protect them is urgent. If you wish to know more contact the Royal Society for Nature Conservation, The Green, Nettleham, Lincoln. On the worldwide scene the toll of plant destruction grows almost by the square mile weekly. Many unknown plants beneficial to agriculture and medicine are lost before their value is known. Comparatively recently a plant, the Rosy Periwinkle, was found in the rain forest of South America and a drug extracted from it has increased the chances of a child recovering from leukemia from 20% to 80%. Despite this, forest destruction goes on. The World Wildlife Fund of Godalming, Surrey, has all the information.

For further information on identification see *The Wild Flowers of Britain and Northern Europe* by Richard and Alastair Fitter, *The Alpine Flowers of Britain and Europe* by Christopher Grey-Wilson, *Food for Free* and *Plants with a purpose* by Richard Mabey: all illustrated by Marjorie Blamey and published by Collins of London.

To help you reach the right part of the book, or at least the right family, a guide is given below. It is not a botanical key but purely a guide in lay terms. For instance, we call most fruits which are globular 'berries' so that is the heading under which you will find them. In the text you will read that a blackberry, for instance, is not a berry but a group of drupes fused together making a compound fruit. Some botanical descriptions may be of help before coming to the guide and are listed below with illustrations.

Achene Small dry indehiscent seed

Berry Fleshy fruit with several seeds without a stony layer

Capsule Dry indehiscent fruit composed of more than one section

Cone Scaly woody structure enclosing seeds between scales

Drupe More or less fleshy fruit with one or more 'stones'

Dehiscent When a fruit splits open to release seed

Indehiscent When it does not

Samara Dry indehiscent fruit with one wing or two

8

Generally the guide leads to a family but where, like the roses, fruit may be of several types those types are listed to guide you to the roses. Pods accurately describe the fruit of the Pea family; also listed under pods are the Cabbage family fruits because they look more like pods than the specialised capsules that they really are. Most of the terms used here are either listed on the previous page or are in common use; however, a pome is a fruit in which the seeds are enclosed by a fleshy receptacle, and a follicle is a dry, one-seeded fruit that consists of a single carpel. Not all types or species are listed, because to do so would take up a disproportionate amount of space. Berries are subdivided and are on pages 10 and 11.

The guide

Achenes – Buttercups, Meadowsweet, Daisies (with pappus)
Capsules – Pinks, Balsams, Willowherbs, Primroses, Plantains, Orchids, Poppies
Cones – Firs, Pines, Hops, Alders, Plane Trees
Discs – Mallows
Feathery – Old Man's Beard, Pasque Flowers, Mountain Avens
Fleshy – Hottentot Fig
Follicles – Columbine, Monkshood
Nutlets – Bedstraws, Figworts
Nuts – Hazel, Beech, Chestnut, Lime, Oaks
Pods – Cabbage, Peas
Pome – Apples, Pears
Samara – Maple, Elm, Ash, Tree of Heaven

Berries

Berries may go through colour changes as they ripen from green to yellow to red or black, for instance. They are listed below under their full ripe colour but may be seen with others in varying stages of ripeness, particularly the Wayfaring Tree.

Black berries

TREES AND TALL SHRUBS	SMALLER SHRUBS	HERBS AND CLIMBERS

TREES AND TALL SHRUBS	SMALLER SHRUBS	HERBS AND CLIMBERS
Juniper	Currant	Berry
Sloe	Dogwood	Catchfly
Bird Cherry	Bilberry	Tutsan
Plum	Privet	Wild Madder
Elder	Oregon Grape	Solomon's
Buckthorn	Blackberry	Seal
	Spurge Laurel	Herb Paris
	Wayfaring Tree	Nightshade
		Ivy

Red berries

TREES AND TALL SHRUBS	SMALLER SHRUBS	HERBS AND CLIMBERS

Yew
Holly
Rowan
Hawthorn
Cherries
Strawberry
  Tree
Guelder Rose

Roses
Raspberry
Cotoneaster
Mezereon
Currants
Dogwood
Cranberry
Butchers Broom

White Bryony
Bittersweet
Black Bryony
Lords and Ladies
Lily of the
  Valley
Honeysuckle

Other colours

Orange: Barberry
  Sea buckthorn
Pink: Spindle

Green: Gooseberry
White: Snowberry
  Mistletoe

Orange: Iris

The first group of plants in the order in which they appear in this book is the conifers. Their fruit is a cone enclosing many seeds, most of which have a papery membrane which enables them to be dispersed by the wind away from the shade of the parent tree.

Silver Fir *Abies alba* (**1**): two white stripes on the underside of the dark green leaves identify this tall tree. The female cones stand erect on the branches, usually near the top of the tree; note the protruding bracts between the scales. When the seed has been shed the empty cones stay like candles on the tree for some months. It is a tree of mountainous regions and was once planted extensively, but the Noble Fir *Abies procera* (not illustrated) has now replaced it. It has larger cones – up to 25 cm – which lose the scales at the same time as the seed, leaving only the thin core on the twigs. **Douglas Fir** *Pseudotsuga menziesii* (**2**) is a very large tree with aromatic leaves growing all round the twigs. As shown, the cones hang from the ends of the branches; note the three-toothed bracts. It was found in North America by the Scottish botanist Archibald Menzies in 1791, and David Douglas, also a Scot, introduced the tree in 1827, giving it 'Menzies' as its specific name. It holds the record as the tallest fir in the world.

12

1

2

13

In contrast to the spiky appearance on page 13 the **Norway Spruce** *Picea abies* (1) has a 'clean shaven' cone. This is the traditional Christmas tree which, if left to grow as a specimen, will attain a height of up to 40 m. The needles are four sided and grow in rows on the twig; typical of the spruces, they leave a scar when removed. The hanging cones have no bracts protruding between the scales. It was introduced into Britain in the 16th century to replace the forests felled for fuel. It is a tree of mountainous country and covers many hillsides, as does the **Sitka Spruce** *Picea sitchensis* (2) which today is planted in preference. These needles are sharply pointed, and note the blue-green underside. Bracts again do not protrude from the cone, and the scales are thin and wrinkled. The seed has a papery wing which takes it in the wind well clear of the parent tree. The timber produced by this tree has long fibres and is soft, making it an ideal plant for the manufacture of paper. The great demand for this material is outgrowing the rate at which the tree can be harvested when grown as a crop. To grow fast and straight it needs a moderate and damp climate, which makes it an invaluable tree for the forest plantations in the western side of Britain.

1

2

15

1a

1b

1c

1d

Western Hemlock *Tsuga heterophylla* (**1**) is a decorative and graceful tree worth planting as a specimen if plenty of room is available; it does not thrive on chalky soils. Note the blunt-ended needles and the whitish appearance underneath. The immature cone (**1a**) at the tip of a shoot ripens and opens its rounded scales (**1b**) to release the seed (**1c**), which rapidly germinates. The empty cones (**1d**) often stay on the tree for long periods before being knocked off by the wind movement. It is usually grown under a nurse crop because it grows best in shelter and in moist climates. Another tree for western areas.

European Larch *Larix decidua* (**2**) is a pine which grows to about 36 m, making a handsome specimen with slightly drooping lower branches. The leaves grow in tufts at all angles with the male flowers (**2a**) and the delightful little miniature female cones (**2b**), which in early spring are sometimes called Larch Roses. Being deciduous the larch allows a good ground flora to develop beneath its spread, unlike the evergreen conifers. Because the **Japanese Larch** *Larix kaempferi* (**3**) is less prone to disease it is more frequently grown. The branches are thicker and do not droop. The two hybridise freely (**4**), showing characteristics of both, which makes identification rather difficult.

2

2a

2b

3

4

17

Western Red Cedar *Thuja plicata* is a handsome tree growing to 25 m, conical in shape and – note – with an erect leader. The scaly leaves are pressed tight to the stem and have a whitish stripe on the underside. They are very aromatic when handled. The bark is a dark purplish-brown and is widely ridged. The empty cones stay on the tree for some time after shedding the seed. The timber is made up of very long fibres which enable water to run off easily and quickly. This property reduces the effects of rotting and makes it an ideal wood for outside work such as cladding tiles, roof shingles, greenhouses and almost any general outdoor joinery. Window frames of red cedar do not need painting – a great saving in house maintenance.

Scots Pine *Pinus sylvestris* once covered vast tracts of Scotland and was the main tree within the Caledonian Forest. The pairs of needle leaves are often twisted. Cones take two years to ripen, so immature ones will be seen at the tip of the branch and the mature ones further up where the tip ended the previous year. The single winged seed is released as the scales open. The tree once provided many masts for sailing ships, and turpentine, tar and charcoal, but now is used more for landscape planting and, when felled, for chipboard and telegraph poles.

19

Lawson's Cypress *Chamaecyparis lawsoniana* (**1**) is conical, growing to 35 m with – note – a drooping leader which distinguishes it immediately from the Western Red Cedar (page 18). When crushed the scale-like leaves give a scent similar to parsley. It is mostly an ornamental tree planted in parks and gardens in some of its many variations. There are golden, silver, yellow and blue-green varieties growing from less than 1 m to more than 15 m. It grows naturally in a small area in America and seed was sent to Messrs Lawson, seed merchants in Edinburgh, in the middle of the 19th century, who named it.

1

Juniper *Juniperis communis* (**2**) is a small tree or shrub. The leaves have a white stripe on the upper surface and are rather prickly; note that they grow in groups of three. It appears naturally in scrub and on heathland throughout Europe. Many varieties have been bred to be suitable for garden and park planting. The green berries mature to the wrinkled blue-black ones in their second year, and it is an oil extracted from these that is used to flavour gin. It was once a valuable medicinal plant. **(P)Yew** *Taxus baccata* (**3**) is a poisonous tree, especially the seed. The scarlet cup enclosing the seed is not poisonous, but is certainly not recommended and should never be eaten. The yew is slow-growing but responds to cutting very well, making it a popular tree for topiary. It is found on limy soil.

21

Walnut *Juglans regia* is a large spreading tree with most attractive bark. There are usually seven leaflets on the aromatic leaves, which are alternate on the branches. The tough green covering of the fruit is pubescent and when ripe opens to release the well-known nut. It is an excellent hardwood tree for the veneers which have been prized by cabinet-makers for hundreds of years. It needs a long hot summer to ripen nuts, so it flourishes better further south than Britain. It was introduced into Britain by the Romans, who brought it with them for the oil it produces. The calorific value of 3,000 per pound makes it one of the richest of all nuts, and it may be eaten raw, pickled or cooked in cakes.

Alder *Alnus glutinosa* is a tree of wet land. The rounded leaves are sometimes notched at the top. The pollen from the male catkins is wind-blown on to the purple female flowers in spring. The previous year's empty cone-like structures remain on the tree throughout the winter. The seed has a corky knob which ensures it floats in the water nearby. Seeds are then blown or taken by the current to a muddy shore, where they germinate. It is a tough wood, often used for clog-making. The leaves and bark yield a natural dye.

Hornbeam *Carpinus betulus* (**1**) grows to about 25 m and has a smooth grey bark furrowing as the years pass. The nuts develop in groups – usually 6-8 within the large leafy three-lobed bracts. The heavier soils of south-east England favour it where it is found in hedgerows and parks. It gets its name from the old English word 'horn' denoting the extreme toughness of the wood, and 'beam' for tree. Mill cogwheels were often made from it, and many butchers' chopping blocks are hornbeam.

1

2a

Hazel *Corylus avellana* (**2**) is a very common hedgerow and scrub tree growing to about 6 m. The nuts develop through the summer within the leafy bracts (**2a**) and in September may be seen together with next year's catkins, which in January lengthen (**2b**) to fertilise the female flower (**2c**) in spring when they hang in profusion from the branches (**2d**). They form an important part of the diet of squirrels, dormice and other wildlife. Harvested during August they will be in good condition (**2e**) for eating or cooking. The **Kentish Cob** or **Philbert** *C. maxima* (**3**) is a strain cultivated from the Balkan Hazel. The wood is pliable and is used for hurdles and baskets; it was the basis of the daub and wattle houses built many centuries ago.

2b

2c

2d   **2e**   **3**

Beech *Fagus sylvatica* is one of the most stately of trees, with a massive trunk covered with a smooth grey bark. The nuts contain 46% oil and are relished by many wild animals and birds. They fall in September or October and attract squirrels, woodmice and jays to the wood. In mediæval times the right to graze beech woods with pigs was granted by owners to local farmers. Little light penetrates to the floor of a beech wood in summer so the flora is usually restricted to early-flowering bluebells and wood anemones. With proper management and clipping it makes an excellent boundary hedge.

Sweet Chestnut *Castanea sativa* is another large tree and has a spirally marked bark. Following a long hot summer in Britain the nuts ripen in October but in normal conditions this is unlikely to happen. The sharp spines on the case are pliable. In summer when the flowers are open they have a rather sweet sickly smell. It is widely used as a coppice tree. The nuts are good to eat raw, roasted, pickled or puréed in a stuffing. Most chestnuts eaten in Britain are imported from France or Italy.

The next four pages concern the Oak family. It is a variable one but all members have typically lobed leaves, with the exception of Holm Oak. They are slow growing and therefore provide a tough close-grained wood. Many live to a considerable age. Identification is most easily checked by comparing leaf and acorn but frequent hybridising between the species might cause difficulties. Few trees provide habitat and food for so many types of wildlife, and with more light penetrating to the floor than in beech woods there will be a good and varied flora. Specimen trees grow into massive spreading plants.

The **Common Oak** *Quercus robur* (**1**) has acorns, in pairs usually, on a long stalk but with sessile leaves.

Durmast or **Sessile Oak** *Q. petraea* (**2**) has acorns in almost stemless groups, but has stalked leaves. Oak seedlings grow quickly from the nut (**2a**). In summer immature acorns (**2b**) may often be found. The shape and size of an acorn (**2c**) helps in identification. Durmast Oak was once the main source of wood for charcoal. So much was being felled in Queen Elizabeth I's reign that she declared this to be undesirable, and another source was to be found (perhaps one of the first laws of conservation). Shipbuilding was another cause of the denuding of many areas of Britain; it would take many centuries of dedicated planting and maintenance to replace the lost trees.

29

These six oaks show some of the variations. **Turkey Oak** *Quercus cerris* (**1**), introduced from Turkey in the 17th century, is increasing in S.E. England as a hedgerow tree; note the sessile cup with its 'mossy' outside. Mature and immature acorns may be seen at the same time since they take two years to ripen. Another oak with slow ripening acorns is the **Red Oak** *Q. borealis* (**2**). In the first year they are almost enclosed in their cup. The leaves colour well in the autumn and this tree is planted in parks as a specimen. It does not tolerate much shade. The whole crown colours at the same time. **Holm Oak** *Q. ilex* (**3**) is an evergreen with leaves reminiscent of holly – hence the *ilex* specific name. The younger leaves are spiny (**3a**) and the more mature upper ones are narrower and untoothed (**3b**). Also shown is the pale downy under surface (**3c**).

In the axils of the veins on the back of the leaves of the **Pin Oak** Q. *palustris* (**4**) note the little tufts of hair. It is not a common tree. It colours well but does not thrive on limy soil, which tends to turn the leaves yellow. The **Cork Oak** Q. *suber* (**5**) is purely an ornamental tree in Britain. The bark is soft and gnarled and the tree is evergreen. In Mediterranean regions the bark is stripped every ten years or so and used for its insulation properties and to make bottle stoppers. It is purely for its ornamental value that the **Scarlet Oak** Q. *coccinea* (**6**) is planted in many parks and gardens. The colour spreads slowly throughout the tree. The small acorns take two years to ripen.

5

6

4

English Elm *Ulmus procera* is a traditional boundary tree and has been planted as such for many centuries, although the ravages of the scolytid beetle and its parasitic fungus are now decimating the species throughout Britain. More than 12 million trees have been killed between 1967 and 1982. The uneven lobes at the base of the leaf do not cover the stalk, and note that the seed is placed towards the notch of the disc. Most English Elms grow from suckers rather than seed, which is rarely viable in Britain. It suckers freely, as can be seen around the stumps of dead trees. Unfortunately these shoots are equally susceptible to disease.

Wych Elm *U. glabra* is smaller than the English Elm and has a more open shape. One of the lobes at the base of the leaf often covers the stalk, and note that the seed is central within the disc. This tree propagates from the seed and is slightly more resistant to Dutch Elm disease, but does succumb in time. It gives a tough and durable timber used for such purposes as keels, groynes and other harbour works. The leaf of the continental form is much wider near the tip than the British form.

London Plane *Platanus hybrida* (**1**), as its name implies, is a hybrid. The brown markings on the leaves are rusty coloured velvety hairs which are lost as the leaf matures. The seed is contained in a 'bobble' that stays on the tree after the seed, which is not viable, is shed. Regular flaking of the bark in large pieces discards the filth of pollution which, makes it an ideal tree for amenity planting in cities. **Sycamore** *Acer pseudoplatanus* (**2**) is another widely planted amenity tree. Note the 90° angle of the winged seeds. A tar spot fungus attacks this species, leaving unpleasant black blotches on the leaves.

3

4

The seeds of maples are in winged pairs called samara, and the angle between the wings sometimes gives a clue to identification. Leaves are palmately lobed on long stalks and usually colour well in autumn. **Norway Maple** *Acer platanoides* (**3**) grows to about 27 m in the wild. A native of Norway, it is hardy and has naturalised in Britain as well as being a good park tree. Many varieties of this species have been bred for garden and ornamental planting, showing different leaf colour. **Field Maple** *A. campestre* (**4**) is smaller growing, not much more than 20 m. Note the rounded lobes and the samara almost in a straight line. It thrives on the chalky soil of south-east England. Many of the finest veneers are maple and it is a tree greatly prized by craftsmen cabinet-makers: Stradivarius used maple for his violins and 'cellos in the 17th century.

Horse Chestnut *Aesculus hippocastanum* is a large spreading tree growing up to 25 m. The bark is smooth and grey and the leaves large palmate with 5-7 leaflets. In spring sticky buds forecast the candelabras of flowering spikes which decorate the tree later. The large brown inedible nut is the well-known conker. Many years ago a game was played where snail shells were used to break an opponent's entry; cob nuts were later used, only to be replaced by the Horse Chestnut. The winning nut was the 'conqueror' which became 'conker' over the years. It is native in south-east Europe. *A. carnea* is a smaller tree with pink flowers.

Holly *Ilex aquifolium* is native to Britain and grows to about 20 m. The leaves are spiny, shiny and dark green above, paler below. The small flowers grow in clusters with male and female on different trees. The berries are poisonous and persist throughout winter if not taken by blackbirds and thrushes. It grows in all areas and habitats which are not actually waterlogged. Holly has always been a symbol of Christmas-tide, and legend has it that it is unlucky to fell a holly tree; nevertheless its dense white wood is good for carving, inlay work and woodcuts – no great disadvantage to the craftsman.

Large-leaved Lime *Tilia platyphyllos* (**1**) has a narrow crown and grows up to 30 m with most of the branches growing upwards. The small clusters of greenish-yellow sweetly scented flowers are much favoured by bees. Long leaf-like bracts grow on the stalk above them. The fruits develop with the five ribbed lobes covered in short hairs. It is probably native to south-west Britain and has been planted elsewhere in town parks and avenues, imparting its delightful scent to the environment. It tolerates hard pruning and is often pollarded, which results in even larger leaves.

1

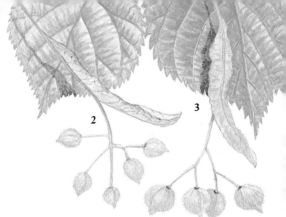

Small-leaved Lime *Tilia cordata* (**2**) spreads more than the large-leaved species, and note that the smooth bark is marked with large bosses on the trunk, which are absent in the other one. Tufts of reddish hairs grow on the back of the leaf, and the long leaf-like bract is at the base of the fruit cluster. The fruit is small and globular. The fruit of the **Common Lime** *T. vulgaris* (**3**) shows characteristics of both the previous species, being a hybrid of them, in that they are large and only slightly ribbed. Many specimens are destroyed annually by various local councils to satisfy demands of car owners. The tree is very prone to attack by aphids and the resulting sticky honeydew which rains down on parked cars makes it an unpopular street tree.

Ash *Fraxinus excelsior* is a large spreading tree, growing to 35 m, with a dark fissured bark. The buds are matt black in the spring before the flowers appear; these look like purplish tufts in small clusters. It is a tree that thrives on limy soil and is hardy. The timber it produces is valuable in that it is hard, white and straight-grained, making it ideal for oars, and axe and other tool handles. It will be most frequently found in hedgerows.

Tree of Heaven *Ailanthus altissima* is an open tree with narrow pinnate leaves, 15-25 leaflets to a leaf. Note the 1-3 lobes at the base of each leaflet. The keys are slightly twisted like those of the ash, enabling them to 'helicopter' their way to earth. It was introduced from China in the 18th century and is often planted in southern city squares and parks. The wood is similar to ash. The seed is rarely viable but the tree rapidly regenerates from suckers, which often appear at some distance from the plant.

(P)Mistletoe *Viscum album* is a multi-branched semi-parasitic plant, dependent usually on apple and poplar trees, but not exclusively so. The flowers are insignificant, four petalled and greenish and produce the single-seeded berries in November. The leaves are tough and rather fleshy. Distribution is by birds who eat the berries with no harm, in spite of their being poisonous to man. They are very sticky (and used abroad as bird-lime to trap birds) and in order to clean their beaks birds wipe them on twigs, thereby planting the seed. In the 17th century a berry was taken from the mistletoe for each kiss either taken or given. No berries – no kisses.

42

Hop *Humulus lupulus* is a square-stemmed climber with hooked bristles, which turns clockwise up its host. The female flowers develop into the cone-like fruit. It is native to lowland Britain and is now cultivated on a very large scale, particularly in south-east England. It is a food plant of the Peacock, Red Admiral and Comma butterflies. Until it was discovered as a flavouring agent for beer various other plants were used, including Ground Ivy *Glechoma hederacea*. It was the Dutch who introduced the idea to British brewers, who to this day use greatly improved strains of the wild plant.

The presence or absence of tubercles on the nutlets, plus leaf shape, helps to identify docks, but beware – they all hybridise with each other. Illustrated below the seed (actual size ×2) is a transverse section (×2) showing the tubercles on the nutlet. Sizes differ widely, so note the scale of the illustrations. **Sorrel** *Rumex acetosa* (**1**) is the only one with edible leaves. There are no tubercles and the lobes on the stemmed leaves point back. **Sheep's Sorrel** *R. acetosella* (**2**) has an almost wingless fruit and leaf lobes pointing forward.

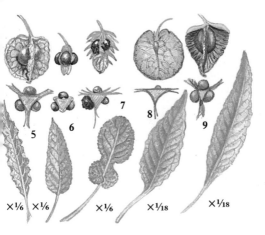

×1/6 ×1/6 ×1/6 ×1/18 ×1/18

Mountain Sorrel *Oxyria digyna* (**3**) has broadly wing-ed fruit and a round leaf. **Broad-leaved Dock** *R. ob-tusifolius* (**4**) has toothed valves with one prominent tubercle. **Curled Dock** *R. crispus* (**5**) usually has all three tubercles swollen, but not always; note the leaves. **Clustered Dock** *R. conglomeratus* (**6**) is well branched and has all three tubercles swollen, but the **Fiddle Dock** *R. pulcher* (**7**) has one only: note leaves and stem. **Northern Dock** *R. longifolius* (**8**) has kidney-shaped fruit and lanceolate leaves. **Water Dock** *R. hydrolapathum* (**9**) is a large plant. All three tubercles are swollen and elongated above the leaves.

Hottentot Fig *Carpobrotus edulis* (**1**) was introduced in the 18th century from South Africa as a garden plant. Like so many strong growing plants it soon outgrew its welcome, and was cast out to become even more vigorous when naturalised. The fruit is edible but is tangy and acid and not recommended; note the 5 fleshy lobes surrounding the fruit. The leaves are triangular and also fleshy. It hangs in curtains down cliffs in the south-west and Wales, and grows equally well on clifftops, taking over large areas. It is of the same family as the well-known and colourful garden plant mesembryanthemum. Another less common species has yellow flowers.

2a

3a

4a

The Pink family (Caryophyllacae) is very widespread, and includes many wild flowers which have been bred for garden planting and for showing. Generally they thrive best in fertile and damp habitats. The campions and catchflies have five white or pink petals above sepals that are joined in a tube to varying degrees. **Red Campion** *Silene dioica* (**2**) has cleft pink petals and pointed leaves. The teeth of the capsule roll back when the seed is ripe (**2a**), unlike the **White Campion** *S. alba* (**3**) on which they stand erect or nearly so. Prior to this they are enclosed (**3a**) by the calyx. The empty capsules are often used as quarters by hibernating ladybirds. The **Bladder Campion** *S. vulgaris* (**4**) has an inflated calyx tube which encloses the seed capsule (**4a**). Quite different in appearance is **Ragged Robin** *Lychnis flos-cuculi* (**5**), with its petals cut so deep that it has a ragged look. The calyx supports a capsule with reflexed tips. This is a plant of wetland.

3   2   4   5

Sea Campion *Silene maritima* (**1**) has much broader petals than the Bladder Campion. Its inflated tube encloses the seed capsule (cut open at **1a**) in the same way. A short plant with some non-flowering shoots, It grows on the coast but may also be found up to 1000 m in mountains. The surprise of the *Silene* group is the **Berry Catchfly** *Cucubalus baccifer* (**2**), in that it is the only member to produce a berry. It is tall with downy, rather brittle stems carrying narrow leaves; the flowers droop (**2a**). Note that the berry retains the calyx of the flower. As it ripens it becomes dull and is much relished by birds. It is naturalised in a few places in south-east England after introduction from Europe.

(P) Corn Cockle *Agrostemma githago* (**3**) is a cornfield weed lost to Britain, but may still be found in land formerly associated with corn. The long calyx teeth (**3a**) hide the capsule as it ripens (**3b**). It is a poisonous weed and was eradicated by hand: cider and buns were once given to those who 'walked the wheat' at Easter, and double shares were given to those who pulled up a plant. **Love in a Mist** *Nigella damescena* (**4**) is mainly a garden plant but might be found in dry open country on chalk. A member of the Buttercup family, it has no petals, but has five blue sepals. The thread-like leaves cling below the 10-celled capsule enclosing the seed (**4a**).

3
×½

4
×½

The flowers of the **Yellow Water Lily** *Nuphar lutea* stand erect out of the still or slow-moving water in which it grows, sometimes to a depth in excess of 2 m. It has a distinct fragrance similar to alcohol, which, combined with the shape of the fruit, has earned it the alternative name of Brandy Bottle. Many leaves float and there are translucent ones just below the surface. The fruits split into segments which in time become waterlogged, sink and germinate in the mud. It is quite distinct from the White Water Lily *Nymphaea alba* (not illustrated), which floats on the surface of the shallower ponds in which it grows.

The **Least Yellow Water Lily** *Nuphar pumila* is smaller with petals more separated than the Yellow Water Lily. It grows in still, even stagnant ponds in the northern upland regions of Europe, but rarely in Britain. It hybridises with the Yellow, both being pollinated by small flies. Hybrids show characteristics of both and are vigorous, like the Fringed Water Lily *Nymphoides peltata* (not illustrated), which is found in ponds and slow- moving rivers of eastern and central England. It is related to the Bogbean and, if planted in ponds, is just as invasive.

Both the Hellebores are poisonous. Much of the mediæval herbal medicine was based on somewhat extreme measures, and these two plants were used as emetics and rather drastic cathartics. Thankfully their use has passed into history.

Ⓟ **Green Hellebore** *Helleborus viridis* (**1**) is a short plant growing in damp woods on limy soil; note that the base leaves die off before winter. The flowers are greenish and open wide, whereas those of Ⓟ **Stinking Hellebore** *H. foetidus* (**2**) hang bell-like in clusters. Note that the leaves are persistent throughout winter. Bees pollinate them and many seeds are distributed by ants attracted by the alaiosome.

Like the previous two species the flowers on this page have no petals but coloured sepals instead. **Marsh Marigold** *Caltha palustris* (3) grows by streams or ponds, even into the mountains. The base leaves are more kidney-shaped than those on the stem, which are heart-shaped. The capsule splits down one side to release the seeds. It is one of the most ancient of native plants in Britain, growing before the last Ice Age. **Winter Aconite** *Eranthis hyemalis* (4) is an introduced species and flowers very early in the spring. The leaves appear after the flowering stem has died down. Three deeply cut palmate leaves grow immediately below the flower, forming a ruff, and persist with the seed capsule.

3   4

Nearly all members of the Buttercup family are poisonous to some degree. Many were used by the old herballists as blistering agents, to relieve gout etc., and as emetics. By modern standards they are highly dangerous when used internally. The seeds are in groups forming heads of various shapes. The most common is the **Meadow Buttercup** Ranunculus acris (**1**). The fruiting head is similar to Bulbous and Creeping Buttercups, R. bulbosus and R. repens (not illustrated) so other parts must be examined. The sepals of this species are erect and the leaves lobed, with the end lobe unstalked. It is found in meadow land up to 1300 m, whereas the **Celery-leaved Buttercup** R. scleratus (**2**) is a lowland plant growing in wetland. The sepals are reflexed and the leaves shiny and palmate. The seed head of **Corn Buttercup** R. arvensis (**3**) is quite unmistakable; the flowers are small with spreading sepals. It grows in damp cornfields, but **Greater Spearwort** R. lingua (**4**), the largest of the group, grows in bogs and fens.

4a
×⅙

Ⓟ**Baneberry** *Actaea spicata* (5) has leaves more like those of an umbellifer, being long-stalked and pinnate. The white flowers are conspicuous in stalked spikes; the berry is highly poisonous and has a strong unpleasant smell. It is not a common plant, but is found on limestone pavements and similar habitats in northern England. It was once used medicinally, after drying and grinding, as a relief for asthma and as a pain killer. It is so poisonous that it is likely it killed more than just the pain.

Old Man's Beard *Clematis vitalba* is a clambering plant, with stems growing up to 30 m long as it covers its supporting plants in hedges in limestone areas. It grips its host with the stalks of its pinnate leaves, helped by tendrils at the end of the stem. The small greenish-white flowers are fragrant. The seeds are in a group, each with a long feathery plume which, en masse, give the plant its vernacular name. The herbalist Gerard named it Traveller's Joy when he saw it first in flower in his home county of Cheshire. It is an unlikely looking member of the Buttercup family.

Another unlikely looking plant of this family is the **Pasque Flower** *Pulsatilla vulgaris*; like the Old Man's Beard it has feathery plumes on the seed. It grows on dry grassland and chalk downs. The basal leaves are in a narrowly segmented rosette. As it ripens the seed heads sometimes droop slightly; the seed heads persist through the summer into the autumn. It is getting rare in Britain as agricultural practices destroy ancient meadows.

×¼

℗ Columbine *Aquilegia vulgaris* (**1**) is a tall plant with long-stemmed root leaves which are trifoliate with rounded leaflets. The papery seed head is made up of 5 follicles which open longitudinally to release the seed. It will be found on scrub and limy woodland. The name is derived from 'columba', Latin for dove: it was said that the flower resembled five doves in a circle. It is another poisonous plant in all its parts. The stems with the seed heads dry well and make good winter decorations, and are very popular with flower arrangers.

×⅙

2

×⅓

3

(P) **Monkshood** *Aconitum napellus* (**2**) is tall and bears its flowers in a long spike above narrowly lobed leaves; note that the follicles are wider than in Columbine. Yet another poisonous plant, it contains aconitum, with which the Greeks and Romans poisoned spears and arrows for hunting. Damp woods and by streams are its natural habitats, and it sometimes occurs in hedgerows as a garden escape. **Wood Anemone** *Anemone nemorosa* (**3**) is a little plant which grows in damp shady places. The flowers have more than six sepals and droop after rain; root leaves appear after the flowers. It contains a narcotic, anemonia, and is poisonous to cattle.

Barberry *Berberis vulgaris* is a shrubby plant growing to about 4 m: note the needle-like spines in groups of three. The berry is edible, but very acid, and useful only when cooked in jams and jellies. It has a very high content of vitamin C and has the herbal reputation of easing digestive problems. In many corn-growing areas it has been eliminated because it is host to a virulent type of rust fungus, but it makes an effective garden hedge and grows naturally in dry woodland.

Oregon Grape *Mahonia aquifolium* is a member of the same family as the rather spiny leaves show: the specific name *aquifolium* is the same as that of Holly. The flowers are very fragrant and appear early in the year. The bloom on the berry is similar to that of the grape, and it is edible, but not very good eating when raw: it makes a good jelly. The wood is fibrous and yellow when cut. It is naturalised in a few places, but will most often be found in parks and gardens.

Poppies have brightened the countryside for many centuries, and have always been a weed of cornfields – the Roman goddess of corn, Ceres, is depicted with a wreath of poppies and corn on her head. They are to be found wherever the earth has been disturbed, because the seed remains viable for many years in the ground. In a modern context, the scar left following the burial of gas or oil pipes across the country is often marked later by a red line of poppies. The seed is small and oily and is distributed through pores around the disc above the seed head, as if from a pepper pot. Numbers 1-4 here exude a white latex from the stem if broken. The **Common Poppy** *Papaver rhoeas* (**1**) is said to have been the original species from which garden varieties originate. Hairs on the main stem stick out at right-angles, making it a roughly hairy plant. The large petals have a black blotch at the base. The pod is never more than twice as long as wide.

On the **Longheaded Poppy** *P. dubium* (**2**) the pod is considerably longer, with a small disc. It is hairless but the hairs on the stem are closely pressed to it, especially near the seed head. **Prickly Poppy** *P. argemone* (**3**) is well named: the stem and leaves are bristly and the ribbed capsule is topped by a small convex disc. The **Rough Poppy** *P. hybridum* (**4**) is distinguished by its round capsule and yellowish bristles. Completely different from all the above is the **Opium Poppy** *P. somniferum* (**5**): the grey-green leaves and lilac flowers contrast with the bright scarlet of the Common Poppy. It is this plant which is the source of the narcotics which are so often abused. The seeds, though, are in no way poisonous, and are used in many culinary ways: they are incorporated into bread and cakes; they are also rich in oil and this is used for salad dressing.

(P)Yellow Horned Poppy *Glaucium flavum* (**1**) looks an unusual form of poppy, with its long seed capsule. The leaves clasp the branching stem and the deeply lobed basal leaves are rough to the touch. The flowers bloom in summer and produce long narrow pods up to 30 cm long. When ripe the capsule splits open, leaving the seeds embedded in the central wall. The stems produce a foul-smelling yellow sap. It is a poisonous plant, said to cause brain damage to those foolish enough to eat it. You will find it growing on coastal shingle and coarse sand; it sometimes grows inland.

1

Welsh Poppy *Meconopsis cambrica* (**2**) is similar to most other poppies, but differs in one important way: the capsule splits open and reflexes to release the seeds. It grows in damp woods and rocky places, and is a frequent garden escape, where it is inclined to be invasive and persistent. **Greater Celandine** *Chelidonium majus* (**3**) is not a celandine. It is about 60 cm tall with branched stems which are brittle and easily broken. It releases a typical orange sap, and it flowers through the summer. The capsule is long and beaded and opens from the base up; the oil gland, or elaiosome, is attractive to ants, which feed on it and so disperse the seeds. The sap is caustic and has been used for many doubtful medicinal purposes in past centuries.

Introduction to the Cabbage family

The Cabbage family (Cruciferae) is one of the very important families of plants used by man, and one of the most confusing to identify because so many hybrids and improved species have naturalised. The name comes from the form in which the flower grows, with four separate petals opposite each other in the shape of a cross. This is not to say that all four-petalled flowers are crucifers; willowherbs and bedstraws, for instance, have the same conformation but are members of their own families.

Many vegetables are the result of selective breeding over the past three or four centuries. The Wild Cabbage is a tough cliff plant, and is the ancestor of many varieties of leaf vegetables. Agricultural fodder crops like kale and rape account for a major proportion of animal greencrop food, without which agricultural production would be sorely tried. It is those plants which escape from the fields which cause identification problems. Some crucifers are delightful garden plants, e.g. Wallflowers and Candytuft, whilst others are tiresome weeds, e.g. that ballistic pest Hairy Bittercress. The smell of woad (with which the ancient Britons are said to have dyed themselves) caused Queen Elizabeth I to declare that production must stop in any areas through which she was to travel: the process involves fermenting the plant, and the stench of rotting cabbage emanating from the field understandably offended the royal nose. The dye was used in those days to colour the uniforms of the police force.

The most reliable clues to identification of many crucifers is the shape of the seed pod and the way in which it grows in relation to the main stem. The flowers are all basically the same and may be white, yellow or pink/violet. Leaves are a great help sometimes, and where relevant are shown.

Typical pods are illustrated below.

. long and narrow

long, narrow and beaded

round, oval or winged

various other shapes or forms.

Wallflower *Cheiranthus cheiri* (**1**) has a woody lower stem and untoothed leaves. It grows in dry limestone areas and on old walls. **Hedge Mustard** *Sisymbrium officinale* (**2**) grows on waste ground and becomes straggly when in fruit (**2a**); the pods open from the base to release the seed (**2b**). The whole plant is rough to the touch and smells of bad eggs if crushed. **Tall Rocket** *S. altissima* (**3**) is well named with its very long pod, and note the leaf shape. The basal rosette of leaves soon dies down. **Creeping Yellowcress** *Rorippa sylvestris* (**4**) as its name implies is stoloniferous and occurs in damp places; note that the pods are the same length as the stalks.

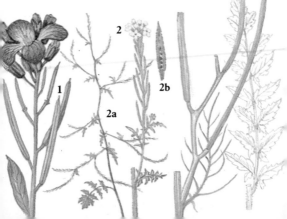

Two plants bred and widely cultivated are the **Wild Turnip** *Brassica rapa* (**5**) and the **Wild Cabbage** *B. oleracea* (**6**). Turnip is a slender plant with leaves that clasp the stem: note how unopened buds appear below the flowers, which are pollinated by bees and hoverflies. By developing the tap root, plant breeders have produced many beets and root crops. By concentrating their attention on the leaves of the Wild Cabbage they have developed most of the leafy vegetables eaten today. This is a robust plant, growing on coastal cliffs in the west country. Note the scars left on the main stem by fallen or harvested leaves, and note that the buds are above the open flowers.

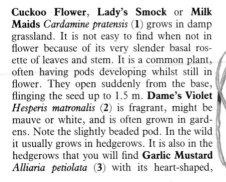

Cuckoo Flower, Lady's Smock or Milk Maids *Cardamine pratensis* (**1**) grows in damp grassland. It is not easy to find when not in flower because of its very slender basal rosette of leaves and stem. It is a common plant, often having pods developing whilst still in flower. They open suddenly from the base, flinging the seed up to 1.5 m. **Dame's Violet** *Hesperis matronalis* (**2**) is fragrant, might be mauve or white, and is often grown in gardens. Note the slightly beaded pod. In the wild it usually grows in hedgerows. It is also in the hedgerows that you will find **Garlic Mustard** *Alliaria petiolata* (**3**) with its heart-shaped, rather nettle-like leaves; its racemes of white flowers show up clearly. Jack-by-the-Hedge is another name for the plant, all parts of which smell and taste of garlic. You will be lucky to find the rare **Tower Cress** *Arabis turrita* (**4**), with its extraordinary pods, in Britain. It is naturalised in walls in the Cambridge area. The seeds have a thin membranous wing. **Watercress** *Nasturtium officinale* (**5**) can be found almost anywhere in slow-moving water. Note the semi-transparent pods, showing the seeds quite clearly. Only eat raw watercress from commercially grown sources, because in the wild it might harbour a parasitic liver fluke, causing severe illness.

3

5

4

Black Mustard *Brassica nigra* (**1**) is a greyish plant with bristly and lobed lower leaves. It is grown as a green manure rather than as a condiment, and is found in hedgerows as a field escape. **Charlock** *Sinapis arvensis* (**2**) is not so obliging and remains a persistent weed. Note the sessile leaves and the immature pod (**2a**) which ripens as **2b**. **Wild Radish** *Raphanus raphanistrum* (**3**) is another weed of arable land. The flowers vary from white to deep yellow with conspicuous purplish veining. The pods snap easily and fall to the ground with the seed (**3a**). In coastal areas from the cliffs to the drift line, **Sea Radish** *R. maritima* (**4**) will be found. Three or sometimes four seeds occur in the heavily beaded pod (**4a**), which does not break

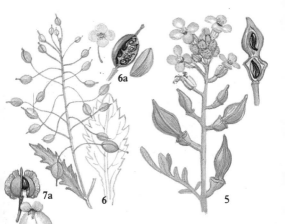

into segments. Pods persist on the plant, often through the winter, and will float in sea water for up to 10 days without becoming less viable, which accounts for those plants found on the drift line. On dunes behind coastal beaches **Sea Rocket** *Cakile maritima* (**5**) will be found with its oval seed pods. It is a succulent plant with an almost crystalline sheen on the petals. **Great Yellowcress** *Rorippa amphibia* (**6**) thrives in wet places and spreads with runners, unlike other crucifers. Note the long-stalked pods containing several seeds (**6a**). **Wild Candytuft** *Iberis amara* (**7**) needs dry limestone ground to thrive. The winged seed pods contain one seed each (**7a**).

Scurvy was once a regular hazard for sailors on long voyages. It was found that by eating a certain coastal plant scurvy was avoided: that plant was **Common Scurvy Grass** *Cochlearia officinalis* (**1**). The leaves loosely clasp the stem and the swollen pod contains up to 6 seeds (**1a**). It grows on dry saltmarshes, its white flowers giving them a shine from the distance. **English Scurvy Grass** *C. anglica* (**2**) is also white-flowered with leaves clasping the stem; the fruit is flattened and clearly veined. Find it on muddy shores, where it often hybridises with **Early Scurvy Grass** *C. danica* (**3**) with stemmed leaves and sometimes pink flowers. Small heads of flowers typify the next few species. **Shepherd's Purse** *Capsella bursa-pastoris* (**4**) is a slender plant growing in most places with its distinctive seed pod reminiscent of the pouches carried by shepherds in the 15th century.

Field Pepperwort *Lepidium campestre* (**5**) has triangular leaves clasping the stem and grows on dry banks, as does **Hoary Cress** *Cardaria draba* (**6**) with its beaked kidney-shaped pods. Small yellow flowers in clusters on stems above hanging pods identify **Woad** *Isatis tinctoria* (**7**). This is an unusual form of seed pod for this family. **Field Pennycress** *Thlaspi arvense* (**8**) has one of the largest seed pods, and will be found on waste ground, but look for **Perennial Honesty** *Lunaria rediviva* (**9**) in damp woods. Note the transparent pods which dry paper-thin, similar to the more often cultivated species *L. annua* (see page 229).

×1/6

2a

2

1

Wild Mignonette *Reseda lutea* (**1**) is a medium-sized branched plant with small pinnate leaves. It is found on disturbed ground, usually on limy soil. It is very faintly fragrant, but nothing to compare with the cultivated garden varieties: at one time it was grown in pots in cities in order to overcome the unpleasant 'odours' of the streets at that time. Derivation of the word seems to be from the Latin *resedo* 'I calm', as it had a reputation as a soporific plant. The idea of the common name being from the French meaning 'little darling' in the 18th century is contradicted by the contemporary name Egyptian Bastard Rocket: the name in fact originates from the Spanish *miñoneta*. The French call it by the Latin name. **Weld** *R. luteola* (**2**) is similar but taller; the seed pod is more globular (**2a**). It is alternatively known as Dyer's Rocket because it provided wool dyers with a good fast yellow dye. **Navelwort** *Umbilicus rupestris* (**3**) is better described by its alternative name Wall Pennywort, since it will be found in wall and rock faces. In the damp areas of western regions it reaches a considerable size. The dead flower often covers the capsule (**3a** and **3b**). In damp and marshy areas you will find **Grass of Parnassus** *Parnassia palustris* (**4**) with its single leaf clasping the stem below its single white flower; it is named following the description by Dioscorides, 2000 years ago, of a plant on Mount Parnassus.

3
×½

3b   **3**

3a

4
×¼

Introduction to the Rose family

The Rose family (Rosaceae) is well known for the scent and colour of the flowers, the flavour and culinary value of the fruits and its attractive herbaceous plants. The fruit of roses develops into a berry called a hip, which is very high in vitamin C and in time of war and emergency has provided all the syrup needed to compensate for deficiency in diet. All roses have fine conspicuous petals of varying colours. Recognition by hips is made easier by reference to the growing habit and the type of prickle. Birds are very attracted to rose hips, particularly after a frost, which makes them more palatable; voles too relish them.

1   1a

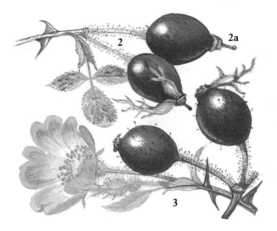

The **Dog Rose** *Rosa canina* (**1**) has long green arching stems in hedges and scrub. Its leaves have 2-3 leaflets and the prickles are strong and curved. The seeds within the hip (**1a**) have sharp hairs which cause severe irritation if eaten, so it is essential to filter through a jelly bag before adding them to jams and jellies or making syrup. *Rosa stylosa* (**2**) is similar but has a bristly stem to the hip. Note the style column (**2a**). The **Downy Rose** *R. tomentosa* (**3**) has downy, paler leaves. This is another with green arching stems and a bristly stalk to the hip, but note that the prickles are usually straighter than in the previous species. Sometimes sepals are persistent.

The **Soft-leaved Rose** *Rosa villosa* (**1**) is an erect shrub.
It has soft downy leaves, blue-green in colour; note that
the prickles are slender and straight. The sepals remain
upright on the top of the hip until it falls and rots. It is
common in woods and scrub in the north, and not very
common in the south. **Sweet Briar** *R. rubiginosa* (**2**) is
also erect with stems covered in hooked prickles and
stout bristles which extend up the stem to the hip. The
sepals usually persist until the hip is fully ripe, and are
sometimes upright. The hips may be hairy or smooth.
The backs of the yellow-green leaves are covered in
spiky brown hairs which give off a sweet apple scent. It
grows mainly on scrub in limy soil.

The **Field Rose** R. *arvensis* (**3**) has weak trailing branches and usually scrambles over other shrubs. The flowers are always white; the prickles are hooked and more or less equal on the stem, which is often tinted red. Note the style column sometimes present on the top of the hip. It is rare in Scotland, becoming common the further south you are. **Burnet Rose** R. *pimpinellifolia* (**4**) is a low-growing shrub, spreading by suckers over large areas. It is covered in straight prickles and bristles. It is most commonly found in coastal areas. Note the dark purple or black hips, usually with persistent sepals.

Botanically the blackberries are one of the most confusing groups to identify, with up to 400 species and sub-species. Hybrids occur frequently, making it even more difficult. Blackberries should be picked before Michaelmas to get the best flavour, after which it is said the Devil throws his cloak over them and they become insipid and watery. They are a favourite autumn fruit of birds, voles and foxes. There are many ways to eat them – raw or cooked with apples as jam or jelly. They make a good wine or cordial and an excellent syrup for flavouring ice-cream. The flowers are usually white but often tinted pink, particularly in buds; the leaves are divided into three, five or seven leaflets and the stems may be angled or almost round. The degree of prickliness varies greatly.

Rubus nessensis (**1**) is an upright plant spreading by suckers. The stem is angled, with short straight prickles on the angles; the three or five leaflets are green on both sides, and note that the fruit is dark red. *R. carpinifolius* (**2**) is a low, arching plant, rooting at the tip with numerous strong yellow prickles on the angles of the stem. The flowers are pinkish in bud and the leaves have five toothed leaflets, hairy and greyish below. *R. ulmifolius* (**3**), a pink-flowered species, is without doubt the most common of the four species illustrated, and has an arched red stem rooting at the tip. It has strong recurved prickles on the angles only. It is roughly hairy all over, with three or five leaflets which are distinctly felted below. It grows in almost any habitat except in deep shade. *R. dasyphyllus* (**4**) is another arching plant with pink flowers and a robust angled red stem with prickles all over. The three or five leaflets are thick and hairy.

Above: berries start hard and green, ripening through red to black.

Dewberry *Rubus caesius* (**1**) has weak round stems growing prostrate. It is less prickly than Blackberry and always has white flowers; note the waxy bloom on the fruit. The leaves are divided into three leaflets. You will find it in damp limy areas. **Raspberry** *R. idaeus* (**2**), on the other hand, grows best in shady areas and heathland throughout Britain. The Raspberry has been recorded since the 17th century as being of culinary and medicinal value. An infusion of its leaves to replace tea for those who cannot tolerate tannin is most acceptable. The fruit freezes well, but is best eaten fresh with cream and sugar. Many hybrid varieties are cultivated.

Cloudberry *R. chamaemorus* (**3**), though, is rarely grown commercially. It is a prostrate plant with rounded lobed leaves. The white flowers are shy of flowering in Britain, so few fruits are found in its natural upland bog habitats. When found it is a very good addition to blackberry pies and jams. **Stone Bramble** *R. saxatilis* (**4**) is prostrate and spreads with runners over shady rocky places. It is spineless and has leaves divided into three leaflets. The sepals are the same length as the petals.

The **Wild Strawberry** *Fragaria vesca* (**1**) is a low scrambling plant in hedges and verges. It roots from long slender runners, enabling it to spread far and fast. Note the pale hairy under-surface of the leaves. The fruit is a berry with the seeds outside; the flavour is good, but their size makes them very fiddly to pick in quantity. They are very high in Vitamin C, and in the 17th century were made into a drink to stimulate the system after a winter's restricted diet low in vitamins. **Herb Bennet** *Geum urbanum* (**2**) or Wood Avens as it is also called, is another plant of the hedgerow. Note the leaflike stipules (**2a**). The hooked seeds are dispersed by being caught in ani-

1

2

2

mals' fur or birds' feathers. In early days the root was used to flavour ale with its clove scent, or hung up to drive away flies. **Water Avens** *G. rivale* (**3**) is found in waterlogged ditches and similar places, more commonly in northern Britain. Seeds are dispersed by the hooks, as with Herb Bennet; the two species hybridise. **Marsh Cinquefoil** *Potentilla palustris* (**4**) has very wide sepals behind the narrow petals. These almost enclose the many seeds within a capsule. You will find it in lime-free wet places.

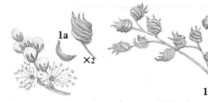

The creamy froth of the flowers of **Meadow-sweet** *Filipendula ulmaria* (**1**) is soon recognised, standing tall in damp grassland. It is said to have been Queen Elizabeth I's favourite strewing herb. Its name derives from its use in the flavouring of mead – the old fermented honey drink. Note the fruiting head made up of several seeds twisted into a spiral (**1a**). You will need a different habitat in order to find **Mountain Avens** *Dryas octopetala* (**2**): this plant thrives in limestone rocky places from sea-level to 2500 m. It creeps with long stems carrying many dark green leaves which are grey on the

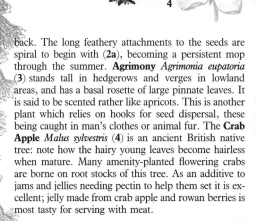

3

4

back. The long feathery attachments to the seeds are spiral to begin with (**2a**), becoming a persistent mop through the summer. **Agrimony** *Agrimonia eupatoria* (**3**) stands tall in hedgerows and verges in lowland areas, and has a basal rosette of large pinnate leaves. It is said to be scented rather like apricots. This is another plant which relies on hooks for seed dispersal, these being caught in man's clothes or animal fur. The **Crab Apple** *Malus sylvestris* (**4**) is an ancient British native tree: note how the hairy young leaves become hairless when mature. Many amenity-planted flowering crabs are borne on root stocks of this tree. As an additive to jams and jellies needing pectin to help them set it is excellent; jelly made from crab apple and rowan berries is most tasty for serving with meat.

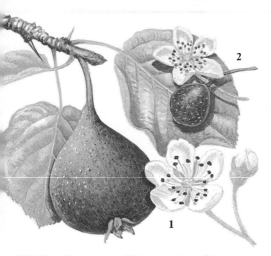

Wild Pear *Pyrus pyraster* (**1**) is not native to Britain and is rarely found. It is spiny with rounded hairless leaves, and in spring has white flowers, clearly distinguishing it from apple. The fruit is woody and dry, and note the sepals retained on the end. The 'Wild Pear' usually seen in hedgerows is the Common Pear *P. communis* (not illustrated), probably planted from discarded cores of cultivated pears, and which have reverted. The twigs are reddish-brown, probably not spiny, with the familiar fruit. The **Plymouth Pear** *P. cordata* (**2**) is very rare in Britain, occurring in the Plymouth district only; it is less rare in western regions of Europe. Note the tiny rounded pear and pink-flushed flowers.

3

Rowan *Sorbus aucuparia* (**3**) grows to about 15 m. It differs from others in the Rose family by having pinnate leaves green on both sides. You will find it in poor soils in mountains and rocky areas; its leaves and habitat give it its alternative name of Mountain Ash. The berries are rich in vitamin C, and when combined with rose hips and crab apples in a jelly are very good to eat with game. Blackbirds and thrushes disperse the seeds. The wood is hard and flexible, which is good for wood carvers, but legends abound and it is said that to fell a Rowan brings bad luck because it is an excellent deterrent against witches and evil spirits.

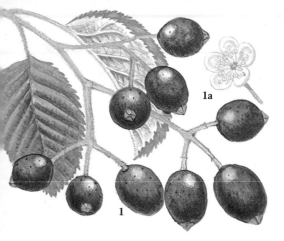

1a

1

Whitebeam *Sorbus aria* (**1**) will be found in copses and on limestone scrubland. Note the pale, almost white underside of the leaves. The flowers (**1a**) are in loose fragrant clusters in midsummer, producing green berries ripening to red in autumn. In the 16th century it was said they were good to eat if they were half-rotten and mixed with wine and honey. The wood is extremely hard and has been used to make mill cogwheels. **Swedish Whitebeam** *S. intermedia* (**2**) is 'half way' between the Rowan and the Whitebeam. The leaves are lobed and yellow-grey on the back; it grows to about 20 m, and is popular with local authorities for amenity planting. Naturalised trees are the result of bird dispersal.

The **Wild Service Tree** *S. torminalis* (**3**), rare in Britain, has deeply lobed leaves: note that the first lobes are approximately at right angles to the stem (**3a**). The leathery-skinned berries are too acid to eat. The name derives, through various changes, from the Old English 'syrfe' to 'serves' in the 16th century, referring to the sorb-apple, as the berries were called in those days.

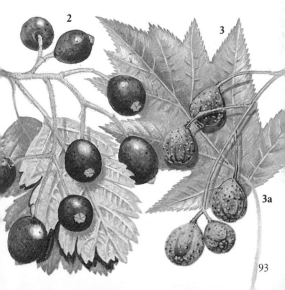

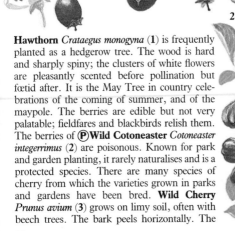

Hawthorn *Crataegus monogyna* (**1**) is frequently planted as a hedgerow tree. The wood is hard and sharply spiny; the clusters of white flowers are pleasantly scented before pollination but foetid after. It is the May Tree in country celebrations of the coming of summer, and of the maypole. The berries are edible but not very palatable; fieldfares and blackbirds relish them. The berries of Ⓟ**Wild Cotoneaster** *Cotoneaster integerrimus* (**2**) are poisonous. Known for park and garden planting, it rarely naturalises and is a protected species. There are many species of cherry from which the varieties grown in parks and gardens have been bred. **Wild Cherry** *Prunus avium* (**3**) grows on limy soil, often with beech trees. The bark peels horizontally. The ancient Greek herbalist Dioscorides recommended the gum from this tree as a cough medicine: note the modern cherry cough mixtures. Note too the calyx constricted at the neck, the smooth-edged sepals (**3a**) and the dark red fruit. **Dwarf Cherry** *P. cerasus* (**4**) is a smaller hedgerow tree. Note here the unconstricted calyx, the notched sepals (**4a**) and the bright red fruit. Check on the drooping leaves of Wild Cherry compared with the more upright ones of Dwarf Cherry. Birds love the small dark cherries of the **Bird Cherry** *P. padus* (**5**). The almond-scented flowers hang in loose clusters; the fruit is very bitter.

1   **2a**

Sloe *Prunus spinosa* (**1**), sometimes called Black-thorn, is a large, very spiny bush with black or very dark brown twigs. The white flowers show clearly against the dark background because they open before the leaves appear. The blue-black sloe is bitter, but frost improves the flavour. It is a tasty addition to fruit pies and is particularly useful in flavouring gin to make the well known liqueur. The fruit needs to be washed, pricked several times and covered in sugar; gin is then added and the whole shaken for 10-14 days. **Bullace** *P. institia* (**2**) is not a thorny shrub. The petals are pure white. The fruit is usually blue-black with a bloom but very rarely may be greenish-yellow (**2a**). It is not grown for its fruit, but the Damson is considered to be a cultivated form of Bullace. **Wild Plum** *P. domestica* (**3**) is a small tree with larger leaves than the other species on these two pages.

2

The fruit is larger too, and is well known in cultivated orchards in various hybrid forms; it cooks and freezes well and makes good wines and liqueurs. Make use of the kernel within the stone for full value. **Cherry Plum** *P. cerasifera* (**4**) is smaller, growing to 6-8 m. The leaves are slightly glossy above. These are four difficult species, because it is possible that (2) and (3) are sub-species of a natural hybrid of (1) and (2). Flesh of (1) and (2) adheres to the stone, whereas in (3) and (4) it is usually free.

Introduction to the Pea family

The next family of major importance to man is the Pea family (Leguminosae). The seeds are contained in pods of different sizes and shapes. Plants within this group grow in many different habitats throughout the world, and vary in size from small creeping herbaceous plants to large and beautiful exotic tropical trees. The flower consists of a large standard petal, two lateral petals – the wings – and two lower ones joined at the bottom – the keel. The standard usually encloses the rest whilst in bud. The leaves are trifoliate or pinnate and sometimes have a terminal tendril.

The economic value of these plants is immense. The clovers and vetches are first rate fodder crops and do well where mixed with grass. Commercially grown peas and beans supply a good proportion of man's needs, providing a vegetable protein-rich diet. These crops are called pulses, and include many varieties. All these plants have nodules on the root stem which are able to store nitrogen taken in by the upper part from the atmosphere. When the plant is eaten, cut down or dies the nitrogen is released into the soil, thus enriching it; consequently its use as a green manure is being stressed to developing countries. Some of the pea-like seeds are very bitter and not worth eating, and the Laburnum Tree produces very dangerous poisonous seeds.

Gorse *Ulex europaeus* (**1**) is a dull green spiny bush growing to about 2 m. The spines are in fact the stiff leaves, each of which is furrowed along its length. The familiar yellow pea flower has a keel shorter than the wings. The scent of Gorse flowers on a hot day has been variously described as being of orange, almond or coconut. It flowers for most of the year, producing seed in dumpy hairy 25 mm pods which explode when ripe, flinging the black seeds several metres. It grows in dry heathland and was once used as a hedgerow plant. **Western Gorse** *U. gallii* (**2**) is very similar to (**1**) and is more common in the west, but only grows to about 60 cm. **Dwarf Gorse** *U. minor* (**3**) is much smaller. The keel is about the same length as the wings and the flowers are denser on the stem. The pods are only about 15 mm and the seeds are olive in colour.

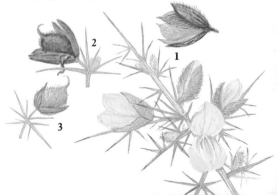

Dyers Greenweed *Genista tinctoria* (**1**) is a small shrub which grows in woodland and on heaths. Note that the standard and keel of the flower are the same length. The pods explode by twisting when ripe, thus dispersing their seed ballistically. The stems produce a yellow dye which was widely used in the 15th century. Cloth dyed thus and then dipped in the blue of woad produced Kendal green, named after the place where this process was developed. **Hairy Greenweed** *G. pilosa* (**2**) is smaller and might even be prostrate on dry heathland. The hairy pods persist on the plant into autumn. **Petty Whin** *G. anglica* (**3**) is a delicate little spiny undershrub. Note that the standard is shorter than the keel. The pods are inflated and twist explosively to scatter the seeds over the dry heathland.

Broom *Cytisus scoparius* (**4**) is a large bushy deciduous shrub growing on heathland, like Gorse. It is spineless with a brighter green, ridged stem. Note the fringe of hairs round the margin of the pods. Bumble bees pollinate the flowers, which are larger than Gorse and unscented. Country people used to make brooms of the leafless branches in winter. **Bladder Senna** *Colutea arborescens* (**5**) has the most inflated pod of this group. The stalked flowers (**5a**) are sometimes marked red. It is a deciduous shrub growing to 3 m in scrub, on railway banks.

Tree Lupin *Lupinus arborea* (**1**) is an upright shrub growing to about 3 m. The pods contain from 8 to 12 black seeds. It was introduced into Britain from California in the 18th century as a garden plant which has since become naturalised on sandy wasteland. It is fragrant and sometimes has white flowers, rarely tinged with blue, showing its relationship to the **Wild Lupin** *L. nootkatensis* (**2**) which is a robust perennial plant, also from America, and introduced at the same time from the Nootka Sound. It has established itself in the shingle banks in some Scottish rivers. The seeds are very bitter and not recommended for eating.

(P) **Laburnum** *Laburnum anagyroides* (**3**) is more often found planted in parks and gardens than naturalised in its normal mountain regions. The cascades of yellow flowers have earned it the name Golden Rain in some areas. The pods hang amidst the dead flowers and contain the black seeds which are scattered explosively. They are the most poisonous of seeds and many children are taken seriously ill after eating them. The wood is hard, pale and fine grained, with very dark heartwood once used as a substitute for ebony.

Vetches are scrambling plants with pea-like flowers and many leaflets on the leaf stalk which ends in a branched or simple tendril. **Tufted Vetch** *Vicia cracca* (**1**) is familiar in hedgerows with its numerous flowers in one-sided spikes. It scrambles up through other plants using the tendrils. Note that the pods are brown.

104

Wood Vetch *V. sylvatica* (**2**) has fewer flowers and leaflets. The flowers are much larger than (1) and clearly veined. Note that the stipules at the base of the leaf stem are toothed and that the pod is black. The flowers of **Bush Vetch** *V. sepium* (**3**) are borne on very short stems from the leaf axils above the stipules which – note – have a dark spot. Note too the blunt end and short spike at the end of the leaf. All species on these pages are perennial and have branched tendrils.

3

Upright Vetch *Vicia orobus* (**1**) is a medium erect plant with pink flowers and pods on a long hairy stem. There is no tendril at the end of the leaf, but note the tiny pointed spike. Its brown pods show up clearly in rocks and scrub. **Spring Vetch** *V. lathroides* (**2**), on the other hand, is prostrate. Note the unbranched tendril and the unmarked stipule. You will most likely find it in coastal sandy areas.

Common Vetch *V. sativa* (**3**) is a hedgerow plant with branched tendrils. Note the marked stipule and the pods usually held nearly upright. The seeds (**3a**) are contained singly in compartments within the pods, which twist open to release them. In a larger form it is sometimes cultivated and the seeds used for pigeon food.

3

3a

Wild Common Vetch *V. angustifolia* (**4**) is very similar to (3) but has narrower leaflets. Both are common plants of hedgerow and grassy places.

4

Slender Tare *Vicia tenuissima* (**1**) is a weak scrambling plant rather than a climber. The tendrils are usually unbranched but may be branched on some plants. The pods are downy and contain 5-8 seeds. It is not a common plant, but you can find it in cornfields and hedges in southern England. Another weak but more common scrambler is the **Smooth Tare** *V. tetrasperma* (**2**). As the specific name implies, the pod contains only four seeds.

Hairy Tare *V. hirsuta* (**3**) is also aptly named. The plant itself is glabrous but the pod, which contains only two seeds, is hairy. As a rule the tendrils are branched, but occasionally are not. Find it in grassy places and on cultivated land.

Yellow Vetch *V. lutea* (**4**) has branched or unbranched tendrils. The very hairy pods contain 4-6 seeds, and note the recurved top.

Bithynian Vetch *V. bithynica* (**5**) is the only vetch with an angled stem, so is more like a true pea. Note the large stipules and wider leaves. The pods ripen to brown and have a prominent curved beak. It grows in grassy places, mainly in western Britain and Europe.

Broad-leaved Everlasting Pea *Lathyrus latifolius* is not native to Britain, but has naturalised itself, often on railway embankments and waste ground. Note the stipules (**a**) in that they are more than half as wide as the winged stem (**b**). The flowers may be white. Both this species and the next are related to the favourite garden Sweet Pea *L. odoratus* (not illustrated) whose flowers vary beautifully in colour.

Narrow-leaved Everlasting Pea *L. sylvestris* is distinguished from the Broad-leaved by the stipules (**a**). Note that they are less than half as wide as the winged stem (**b**). Check too on the colour of the flowers. In this plant they are rather muddy and they are smaller. Both species are climbers through neighbouring plants.

Meadow Vetchling *Lathyrus pratensis* (**1**) scrambles with square stems in hedgerows or in grassland. The flowers and pods are borne on stems longer than the leaves, and are in groups varying from 4 to 12. Note the prominent stipules (**1a**). The tendril may be forked but is usually simple; the hairy pods explode to scatter the seeds. It has a creeping rootstock, as does **Bitter Vetchling** *L. montànus* (**2**) which has a tuberous root that used to be stored, dried and eaten. In Scotland it was once used to flavour whisky with its chestnutty fragrance. It has winged stems and grows in hedgerows, thickets and woodland. The flowers fade to various shades of blue and almost green.

The flowers of the **Grass Vetchling** *L. nissolia* (**3**) also fade in this way. The leaves are in fact modified stems, becoming grasslike: this makes it very difficult to find in its grassland habitat. The pods ripen to a pale brown. **Hairy Vetchling** *L. hirsutus* (**4**) also fades to a blue/green with a creamy keel. Its downy pod is most noticeable, but you would be lucky to find this rare vetchling in its wasteland habitat.

Marsh Pea *Lathyrus palustris* (**1**) is a scrambling plant, slightly downy, with a winged stem. The flower stem is longer than that of the leaves, and the stipules at the base are narrow and arrow-shaped. The pods contain 3-8 seeds which ripen black. It is a rare plant, occurring in a few marshy localities. The **Yellow Vetchling** *L. aphaca* (**2**) is a very rare plant of sandy or gravelly places on chalk. Note that the leaves are merely long unbranched tendrils with very wide triangular stipules at the base, clasping the stem.

Sainfoin *Onobrychis viciifolia* (**3**) is a valuable plant cultivated for cattle fodder. There is a terminal leaflet on the leaves. It is many-branched, and carries the one-seeded pods in spikes; the pods do not open to release the seed, but when ripe fall complete to earth. It naturalises in areas in which it has been cultivated. **Spiny Rest-harrow** *Ononis spinosa* (**4**) is a small erect shrubby plant. There are two lines of hairs on the stem. It has sharp spines, unlike the similar Rest-harrow *O. repens* (not illustrated). This plant is hairy all round the stem, and creeps with rooting nodes. Both plants grow on grassy wasteland.

3

4

115

Kidney Vetch *Anthyllis vulneraria* (**1**) is a low plant with pinnate leaves. The flower head is divided into 2, 3, or 4 groups, each with a 'ruff' of bracts. The colour varies from yellow to reddish, sometimes on one plant. The calyx is silky and inflated, and wraps around the single seed per floret. Like many other hairy plants it was once used to staunch bleeding wounds. It grows on chalkland, often on cliffs, and is a food plant of the Small Blue butterfly.

Birdsfoot Trefoil *Lotus corniculatus* (**2**) is very common in rough grassland. It is almost prostrate in habit, with erect flower stalks. There are five leaflets, two of which are separate and look like stipules. Sometimes the yellow flowers are tinged with red, giving them an orange colour. Note how the pods open in a twisted fashion: by doing so suddenly they eject the seed with some force. **Narrow-leaved Birdsfoot Trefoil** *L. tenuis* (**3**) is very similar but has fewer pods per head. The stem is more wiry, branched and erect. It is more common in the south. Both plants feed the Green Hair Streak and Dingy Skipper caterpillars.

Greater Birdsfoot Trefoil *Lotus uliginosus* (**1**) is larger than the previous two species and is more erect, growing to medium stature. Note the greyish underside of the leaves. There may be up to 12 flowers in each head and the plant may be hairy or glabrous; it grows in wet grassland, but rarely in the north. The **Hairy Birdsfoot Trefoil** *L. subbiflorus* (*hispidus*) (**2**) is prostrate and has leaves similar to *L. corniculatus*. Note the long flower stalk and the small pods only 6-12 mm long. It is hairy all over and grows in grassland, most frequently in the south-west of England and in western Europe.

Slender Birdsfoot Trefoil *L. angustissimus* (**3**) has only 1-2 flowers per head on stalks shorter than those of the leaves. It grows from a central stock, from which the stalks spread outwards. Note the very long pods – up to 30 mm – single or in pairs about this hairy plant. Find it in southern coastal regions of Britain and western and southern Europe. The **Horseshoe Vetch** *Hippocrepis comosa* (**4**) is well named, its pod segments being horseshoe-shaped and linked together. There are 5-8 flowers in the head on long stalks. It is prostrate and grows from a central stock in grassy places on chalk.

Birdsfoot *Ornithopus perpusillus* (**1**) is a prostrate slender annual with pinnate leaves: note the small flowers with red veining, and the beaded pods. Each segment contains one seed and breaks off the pod as it ripens. It grows in dry grassland and on bare places, mainly in the south. **Purple Milk Vetch** *Astragulus danicus* (**2**) has many pinnate leaves growing from a branched root stock. The flowers are borne well clear of the leaves (**2a**); the pods are covered with short white hairs and contain up to 7 seeds. It grows on calcerous ground in grass, mainly in eastern Britain.

Milk Vetch *A. glycyphyllos* (**3**) is found throughout Britain, but since it grows straggly in long grass it is easily overlooked. The milk reference is to its reputed beneficial effect on the milk production of goats; its alternative name of Wild Liquorice has no connection with the confection of that name. (Liquorice comes from a related plant, and is the condensed juice of the root pulp.) The word *glycyphyllos* can be translated as 'sweet leaf', so perhaps it was once used as a sweetmeat. Open the pods and you will see they are divided lengthwise, with seeds on each side of a central membrane.

3

Crown Vetch *Coronilla varia* (**1**) is a scrambling plant, often seen in large patches on grassy road verges. It might be white, pink or purple. If controlled it makes a good garden plant as a base for shrubs. Up to 20 flowers might make up the head; immature pods (**1a**) may be seen whilst the plant is in flower. The seeds are contained in waisted segments (**1b**) of the pod. It is an introduced species in Britain, and is native of central and southern Europe.

Lucerne *Medicago sativa* (**2**) is one of the most valuable of fodder plants for cattle. Known as Alfalfa in America, it yields a large tonnage of good-quality food. The flowers are in varying shades of violet and give way to the pods as they develop into spirals with 2 or 3 turns. If seeds are placed in a jar and moistened daily they will soon sprout: these can be eaten, and make a good nutty addition to salads. **Sickle Medick** *M. falcata* (**3**) is similar, but with yellow flowers and sickle-shaped pods. Note that the leaves are toothed at the tip. East Anglian grassland is the most likely place to find it. These hybridise and produce a plants showing characteristics of each – *M.* × *varia* (**4**). Flowers vary in shade and pods may be almost straight or slightly spiral.

Black Medick *Medicago lupulina* (**1**) has a small compact head containing up to 50 tiny flowers. They are usually self-pollinated and ripen to black coiled pods (**1a**); these contain only one seed. Note that each leaflet terminates in a very small point at the end of the central vein. It is common in grassy places, and is cultivated as Nonsuch. **Spotted Medick** *M. arabica* (**2**) is clearly identified by the black spot on the leaves. This seed pod too is coiled, but does not ripen black. It is spiny and about 5 mm across. It grows on base ground and in grassy wasteland. The three Melilots on the next page are rather similar, but there is sufficient difference to enable identification. Like several other plants, notably Woodruff, they contain coumaria, a substance which produces the scent of drying new-mown hay. **Ribbed Melilot** *Melilotus officinalis* (**3**) has flowers with wings and standards of the same length, and a short keel producing hairless brown pods. **Tall Melilot** *M. altissima* (**4**) has all three flower parts the same length, and downy black pods. **White Melilot** *M. alba* (**5**) has wings and keel equal, and both shorter than the standard and the hairless brown-veined pods. Pods and flowers may often be seen at the same time. The plant at (3a) is shown half actual size.

3a
×½

3

4

5

The addition of clovers to grass seed mixtures for agricultural use greatly increases the food value of the crops. **White Clover** *Trifolium repens* (**1**) is most commonly used and is an early nectar-producing plant for bees. Like most other clovers it does not produce a very noticeable fruiting head. The flowers droop after pollination, covering the developing pod. Some species have more distinctive fruit. In **Hop Trefoil** *T. campestre* (**2**) the flowers which cover the pods dry to a russet brown. The calyx tubes of the **Strawberry Clover** *T. fragiferum* (**3**) enlarge, making the head look like a pale strawberry.

Haresfoot Clover *T. arvense* (**4**) thrives in coastal sandy soil. The elongated seed heads are covered with long hairs. **Fenugreek** *T. ornithipodoides* (**5**) is a slender prostrate plant with 2-5 flowers in the head. Note the relatively large pods containing 5-8 seeds. **Subterranean Clover** *T. subterraneum* (**6**) has the ability to bury its pollinated seed head in the sandy soil in which it grows. Both immature (**6a**) and mature fruits (**6b**) are shown.

4

5

6

6a

6b

Perennial Flax *Linum perenne* (**1**) is a grassland plant
more commonly found in the western parts. The
sepals are blunt, whereas those of Pale Flax *L. bienne*
(not illustrated) are pointed. Linen has been made of
flax since the Pharoahs of Egypt; it is now cultivated
for the oil in the seed – linseed – used for animal
food, putty and varnishes. **Musk Storksbill** *Erodium
moschatum* (**2**) smells musky and is stickily hairy with
broad leaves.

Common Storksbill *E. cicutarium* (**3**) has narrower lobes and a black spot on the upper petals. Both grow in coastal sandy areas into which the seeds are 'screwed' by hydroscopic action. Cranesbills, on the other hand, disperse ballistically. The styles and anthers are fused together and remain prominent and erect. As the seed ripens tension builds up until the style suddenly splits open from the base, flinging the seeds up to 3 m. Below (**4**) are four stages of seed production of **Meadow Cranesbill** *Geranium pratense*.

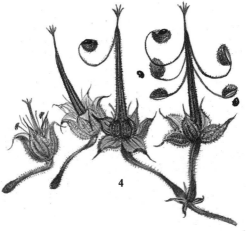

4

(P) **Caper Spurge** *Euphorbia lathyrus* is rare in Britain. It is a plant of bare wasteland. It was once used as a laxative, but being poisonous it killed more than it cured. Gerard warned against it with the old proverb 'Deare is the honie that is likt out of thornes'. The caper-like seed has been used as a substitute in caper sauce – a dangerous practice causing severe internal burning. It is reputed to discourage moles, but with variable results.

×⅑

Himalayan Balsam *Impatiens glandulifera* is another plant introduced to Britain in the 19th century as a garden plant, only to be dismissed when its invasive habits became evident. The flowers vary in shades of pink to a deep reddish colour. The plant (**a**) is fleshy and produces pods (**b**) which, when ripe, suddenly explode (**c**), casting the seeds many metres. Other species of varying colours grow similarly in wetland.

(P) **Mezereon** *Daphne mezereum* (**1**) flowers in the axils of last year's leaves. It is delightfully fragrant and often cultivated for its scent. It has no petals but sepals joined in a tube, at the base of which the red berry develops. It is extremely poisonous. Introduced in the 17th century, it was first reported as being naturalised in 1752. (P) **Spurge Laurel** *D. laureola* (**2**), on the other hand, is native to Britain and grows in woodland on calcareous soil, as does (**1**). The clusters of green-yellow flowers, which develop into black shiny berries, are slightly scented, and open during very early spring before the woodland canopy develops and shades the sun.

1

2

Because it is extremely poisonous, washing after handling any part of the plant is recommended. It is evergreen, but it is neither a laurel, whose leaves it resembles, nor spurge, whose flowers some find reminiscent. It was once used as a treatment for cancer with unrecorded results – its poison no doubt hastened the end.

Shrubs with inconspicuous flowers producing among the most popular fruits follow the two previous highly poisonous ones. The **Red Currant** *Ribes rubrum* (**1**) hangs its fruit in loose drooping spikes. The leaves are not fragrant. It is native, but most plants are garden escapes; its native habitat is damp woods and hedgerows. The berries are acid in taste and rather 'pippy'. Best made into a jelly, and if combined with rowan, rosehips and crab-apple makes an extremely good accompaniment to meat. They add a tang to raspberries eaten with cream and sugar. **Mountain Currant** *R. alpinum* (**2**), though, is insipid and not worth eating. It too grows in woodlands but in those of upland regions. Note that the fruit spikes are erect.

134

The next two plants are both native to Britain but are more frequently found as garden escapes; both are widely cultivated. **Black Currant** *R. nigrum* (**3**) has similar leaves to (1) but they are pungently fragrant if crushed. It grows naturally in damp woodland. The fruit is very vitamin-rich, which makes it an excellent syrup drink for children. Black currant tarts and jams are easily made, but the presence of the persistent calyx makes 'topping and tailing' a necessary tedious job. The same can be said of **Gooseberry** *R. uva-crispa* (**4**). This thorny shrub too grows in damp woodland. It is best cooked since it is rather acid if eaten raw. Both these fruits freeze well.

(P) Spindle *Eunonymus europaeus* grows to about 6 m in hedges and on limy scrubland. Its leaves colour brilliantly in autumn. The flowers grow in small clusters and are insignificant. The 4-lobed, coral-pink berries open in late winter (**a**) to expose the orange seed, making it one of nature's daring clashes; the berries are poisonous and not attractive to birds. The close-grained wood is ideal for pegs, skewers and spindles used for spinning wool in many parts of the world. A mediæval delousing powder was made of dried spindle berries.

Another poisonous-berried shrub is Ⓟ **Buckthorn** *Rhammus catharticus*. Male (**a**) and female (**b**) flowers grow on separate trees. It is a food plant of the Brimstone butterfly, and birds take this berry, which is extremely purging in man. The colour Sap Green was originally made from the berries.

(P) Alder Buckthorn *Frangula alnus* grows to 4 m in wet peaty conditions. The inconspicuous flowers grow at the base of a leaf stalk and develop into a green berry, which ripens through red to a purple-black. It is poisonous and causes severe vomiting if eaten. A greenish-blue dye can be made from the berries, and the close-grained wood makes good skewers. The best charcoal included in gunpowders is made from this wood.

Sea Buckthorn *Hippophaë rhamnoides* suckers freely, making impenetrable thickets in coastal dunes and by high moorland rivers. It first colonised such areas 10,000 years ago when the ice retreated. The sour berries, very rich in vitamins A and C, remain on the trees throughout winter. Thrushes and, particularly, fieldfares relish the berries on the east coast after their long flight from Scandinavia during their winter migration. The thickets make good refuges for many species of wildlife.

All the plants on these pages are soft to the touch. **Musk Mallow** *Malva moschata* (**1**) has deeply cut leaves and is generally hairy. Note that the nutlet (**1a**) turns black when ripe, and that each one contains just one seed. Its softly hairy leaves were once used to staunch bleeding. The **Common Mallow** *M. sylvestris* (**2**) is sprawling and has larger lobed leaves. The fruit is made up of sharply angled nutlets (**2a**) containing brownish-green seed. Both plants contain much mucilage and have been used as poultices in the past. They were once eaten as a vegetable.

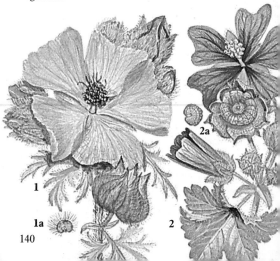

3   4

Tree Mallow *Lavatera arborea* (**3**) is similar to the previous species but much bigger. You will often find that the flowers and maturing fruit occur on the plant together, and that having flowered it turns light brown and has the look and feel of chamois leather. It is more commonly found in western coastal regions. **Marsh Mallow** *Althaea officinalis* (**4**) has velvety lobed leaves folded like a fan. The nutlets are browny-green and softly hairy. Due to their high content of mucilage – a glutinous secretion – they have been used medicinally and in the confectionary trade. This close relative of the garden Hollyhock grows in brackish marshes in coastal areas.

141

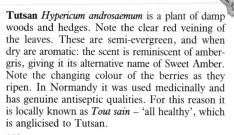

Tutsan *Hypericum androsaemum* is a plant of damp woods and hedges. Note the clear red veining of the leaves. These are semi-evergreen, and when dry are aromatic: the scent is reminiscent of ambergris, giving it its alternative name of Sweet Amber. Note the changing colour of the berries as they ripen. In Normandy it was used medicinally and has genuine antiseptic qualities. For this reason it is locally known as *Tout sain* – 'all healthy', which is anglicised to Tutsan.

142

(P)White Bryony *Bryonia cretica* is the only member of the Gourd family to grow wild in Britain. Using tendrils it clambers over hedgerow plants to about 4 m. Male (**a**) and female (**b**) flowers grow on separate plants. The berries are dull red and are persistent through most of the winter. They are extremely poisonous, and only a small number is sufficient to kill a child. Rabbits will not graze this plant, and it is often found near large warrens for this reason.

Enchanter's Nightshade *Circaea lutetiana* (**1**) uses hooks to disperse its seed (**1a**) by catching in the fur of animals. It grows in shade, and also spreads vegetatively by its whitish stolons. **Large-flowered Evening Primrose** *Oenothera erythrosepala* (**2**) is a Victorian garden introduction which has naturalised. Note the red-tipped developing capsule (**2a**) with the seed neatly packed inside (**2b** and **2c**). The empty seed heads become woody and persist through the winter. Its native American habitat is sand dunes and it thrives under these dry conditions in railway embankments and wasteland. The seed can remain viable for 40 years.

2c

2b

1

1a
×2

2
×¹⁄₁₀

2a

3
×1/16

4

Willowherbs have many things in common. The flowers are all pink or mauve-pink with four notched petals and eight stamens. The leaves are lanceolate – except one – and look like the tree family after which they are named. They can spread vegetatively, smothering out all competing species by forming dense clusters. The seed capsules are long and develop below the flower – botanically speaking they have 'inferior ovaries' – and split open from the top to release the seeds complete with long silky plumes which enable them to be wafted away on the slightest breeze. **Rosebay Willowherb** *Epilobium angustifolia* (**3**) is the tallest, and the small prostrate **New Zealand Willowherb** *E. brunnescens* (**4**) is the smallest, and is the exception, with rounded leaves.

145

Cornelian Cherry *Cornus mas* (**1**) is a shrub of woods and scrub. It is often planted for its attractive early flowers borne on bare yellow-green twigs. The fruit is a drupe and sometimes used to make wine in France – 'vin de cornouille'. It also makes a rather sweet jelly. **Dwarf Cornel** *C. suecica* (**2**) is quite different, being a short creeping perennial. It is a shy late summer flowerer on heaths and high moors. When flowering the large white bracts behind the purplish-black flowers make them conspicuous. The drupe is sweet and not very good to eat, and is best left for wild creatures to harvest: it is shown cut open.

Dogwood *C. sanguinea* (**3**) is a deciduous shrub immediately recognised by its bright red wood. It is for this value that it is grown in parks and gardens. When naturalised in scrub and limy hedgerows the clusters of black berries are evident. They are bitter and not good to eat. When grown for the splendour of the red twigs the plant should be cut down severely in autumn. At one time butchers' skewers were made of this wood, and then they were called 'dogs', which perhaps gave it its vernacular name.

(P)Ivy *Hedera helix* will grow up a tree or wall or creep on the ground with its clinging roots. The leaves on non-flowering shoots are 3 or 5 lobed, shiny and sometimes tinged with mauve on the back. Those on flowering shoots are triangular with wavy edges. The poisonous berries are a favourite food of pigeons, and the shaded dry habitat within the plant is ideal for many insects. The Brimstone butterfly hibernates here. Ivy is much maligned as a strangler of trees, but it is often a supporter of dead or dying ones. Legends abound, and it is said that the ivy decorations in a house at Christmas time protect the occupants from the mischief of goblins who are particularly active at the festive season.

Umbellifers are plants which have their flowers in a flat-topped umbel supported on stems like the spokes of an umbrella. The typical small flowers (**1**) are often present with the maturing seeds which will turn brown or black (**2**,**3**,**4**) depending on the species. The dead plants become woody and often persist through the winter (**5**) and make good indoor decorations. To identify species check on the maturing head (**6**) and the presence of bracteoles (**7**) or bracts (**8**). Finally note the fruit shape as it varies from round (**a**), oval (**b**), bristly (**c**), winged (**d**), oblong (**e**) to very long and narrow (**f**). This group is arranged in this order on the following pages, with white flowers followed by yellow ones. To clarify the shape a single fruit is shown twice actual size with a transverse section of a mericarp (a 1-seeded section of an ovary with more than one section fused together).

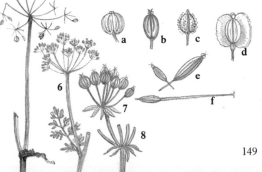

Sea Holly *Eryngium maritimum* (**1**) grows in coastal sand and shingle. The leaves have thickened edges and are sharply spiny. The whole plant is glabrous, which enables it to withstand its often harsh habitat. Bristles on the seed capsule aid distribution. It has been removed from many popular beaches and is now scarce in some areas. It makes a very good garden plant, giving welcome colour in the autumn.

Fools Watercress *Apium nodiflorum* (**2**) grows in wetland, is more or less prostrate, and has been taken for watercress. It is not good to eat – great care should be exercised when looking for real watercress. It spreads rapidly from stem roots at lower leaf junctions. The fruit is round (**2a**) and five-sided (**2b**). Note that there are no bracts.

2a
×4

2b
×4

2

3a

3b
×2

3

Coriander *Coriandrum sativa* (**3**) is a short plant with a fœtid smell. Note the unequal length of the pink or white petals. The seeds remain as a cluster (**3a**) and when ripe become aromatic (**3b**). They are an essential ingredient spice for curry and make a pleasant addition to milk puddings.

151

(P)Hemlock *Conium maculatum* (**1**) is a very poisonous plant in all its parts. Note the hollow stem, purple blotches and glabrous fleshy feel. Bracts and bracteoles are present on the outer side. The seeds are wavy, ridged and round. It is said that Socrates died of a dose of coniine, the poison contained in the plant, which was the official way out of politics in those days.

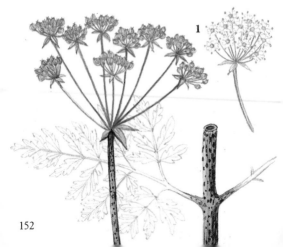

Ground Elder *Aegopodium podagraria* (**2**) has oval fruit, no bracts, and white flowers. Note the persistent styles. It is an invasive weed in gardens, introduced to Europe by the Romans as a cure for gout. It is a plant of damp shady places.

×2

2

Burnet Saxifrage *Pimpinella saxifraga* (**3**) seed is a little more egg-shaped, brown and shiny when ripe. There are no bracts, and note the difference between the upper and lower leaves. It grows on dry grassy places on limy soil. The roots smell of billy goats.

×2

3

153

Greater Burnet Saxifrage *Pimpinella major* (**1**) is similar to Burnet Saxifrage but has larger leaflets. The fruit is oval and no bracts are present. It grows in shady places.

×2

1

Spignel *Meum athamanticum* (**2**) only has bracteoles below the oval fruit. Note that the leaflets are thread-like, giving it a feathery look. It is very aromatic and will be found in mountain grassland. (P)**Fools Parsley** *Aethusia cynapium* (**3**), on the other hand, is a weed of cultivation. It is a poisonous plant and on no account should be eaten. Note the very ridged fruit and the long bracteoles hanging down. It has in the words of the 16th century John Gerard 'a naughtie smell'. Both plants, shown on the opposite page, have prominent sheathing at the base of the petiole.

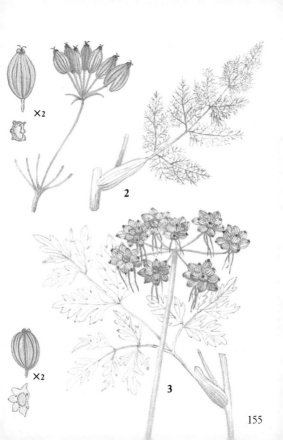

×2

2

×2

3

Greater Water Parsnip *Sium latifolium* (**1**) has many bracts and bracteoles below oval ridged fruits (**1a**). The stem is hollow and grooved. It will be found in fens, wet ditches and most wetlands. The short styles are persistent.

Corn Parsley *Petroselinum segetum* (**2**) grows in dry places and smells of the well-known garden species. It is more commonly found in western areas. Note the irregular or uneven umbels and the long and narrow bracts and bracteoles.

Sanicle *Sanicula europaea* (**3**) is a medium plant growing in beech and oak woods on lime. The pink or white flowers are followed by bristly fruit. It has a healing history from the 15th century, when it was included in a drink with Bugle and Yarrow. 'Bugle holdith the wound open, the Mylfoyle clensith the wound and Sanycle helith it' is how it was described at the time.

Upright Hedge Parsley *Torilis japonica* (**4**) has a solid, roughly-haired stem. It has bracts and bracteoles below the hooked bristly fruit. **Spreading Bur Parsley** *T. arvensis* (**5**) has curved bristles and hairy bracteoles whereas **Knotted Bur Parsley** *T. nodosa* (**6**) has no bracts and only one side of the fruit has bristles, which are rarely hooked.

Wild Carrot *Daucus carota* is a medium-sized plant with a dense umbel of white flowers: note the one single red floret, usually growing in the centre. The bracts and bracteoles are conspicuous and forked. As the fruiting head ripens it becomes concave. It is a plant of coastal grassland and the hooked seeds are dispersed in the coats of wild creatures. It is not the plant from which the cultivated carrot *D. sativa* was bred.

×2

Milk Parsley *Peucedanum palustre* grows tall in marshland. It yields a milky fluid if broken when young. The bracts are variable in number from none to many. It is the food plant of the very rare – in Britain –Swallowtail Butterfly.

×2

Angelica *Angelica sylvestris* is tall, finely hairy at the top, usually with a purplish stem. The umbel is globular with pink or white flowers and the fruit is flattened and winged for wind dispersal. Note the inflated sheath clasping the stem. Bracteoles are very narrow and bracts are usually absent. You will find it in damp woods and grassland.

The **Garden Angelica** *A. archangelica* is naturalised in many areas, often by the sea. The stem is green and the flowers greenish-white. The fruit is winged. It is the side stems (**a**) that are used in the confectionery trade for candy, and by the French in preparing the liqueur Green Chartreuse. It was known in the Middle Ages as a valuable medicinal plant, particularly in relieving digestive problems. It was introduced into Britain in the 16th century.

×1/9

Hogweed *Heracleum sphondylium* (**1**) is a tall sturdy plant found in open woodland and in grassy places. The flowers are white, often turning pink as they mature. Note that those on the outer edge of the umbel are larger and with no bracts as a rule. The mature seeds have a slight wing (see the section diagram). It is related to the huge Giant Hogweed *H. mantegazzianum* (not illustrated). Both plants have serious irritant compounds relative to their size, and should not be touched by those with delicate skin.

×¹⁄₁₀

Cow Parsley *Anthriscus sylvestris* (**2**) is a common roadside plant flowering in sequence with two other parsley-like plants: Cow Parsley from April to June, Rough Chervil from June to July and Upright Hedge Parsley from July to September. Its fruit turns black when ripe and no bracts are present.

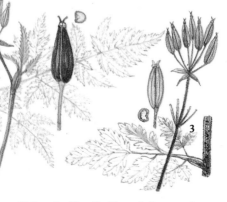

℗ Rough Chervil *Chaerophyllum temulentum* (**3**) is similar but has rough purple spotted stems. It is slightly poisonous and is a common roadside plant.

163

Caraway *Carum carvi* (**1**), a rather rare plant in Britain. It has finely pinnate leaves. The oblong ridged fruit is used as an aromatic flavouring in cakes and the liqueur kümmel. It was first mentioned on a papyrus of the period 2500 BC. Bracts may or may not be present.

Scots Lovage *Ligusticum scoticum* (**2**) grows on rocky coasts in northern Britain. Bracteoles and bracts vary from 1 to 5 and are linear. Note the magenta markings on the lower part of the plant. The leaves can be eaten as a pot herb and a cordial is made to flavour drinks.

Pignut *Conopodium majus* (**3**) has small stem leaves finely cut with a slightly inflated base. The styles remain upright on the fruit, which is oblong. There are a few bracts. The nut-like swollen tubers were once a staple food, and are good to eat if prepared in a broth. It grows in open shady woodland. **Golden Chervil** *Chaerophyllum aureum* (**4**) has yellow aromatic leaves. It looks like Cow Parsley but is stouter and has solid stems. Both bracts and bracteoles are present. You will find it in meadowland in the north.

Sweet Cicely *Myrrhis odorata* is a tall plant common in the north, particularly near habitation, because of its historical use as a fragrant herb. The long-ribbed ripe fruit is shown in transverse section. The leaves give the plant a feathery look. The aniseed scent is used to culinary advantage with the sweetness of the leaf. Place some in the bottom of a saucepan when cooking fruit: it saves using sugar and imparts a pleasant flavour. It is particularly good with rhubarb.

Shepherds Needle *Scandix pecten-veneris* is well named, after the very long fruit beak. The umbel is small, with shorter stalks above broad-toothed bracteoles. Once a weed of arable land, it is now found only in tracks and wasteland.

Alexanders *Smyrnium olusatrum* is one of the earliest coastal plants to make its presence known. It has very swollen sheaths around the main stem at the base of the upper leaves (**a**). The flowers and fruits may be found together (**b**). The fruits ultimately develop into black globular segments which are very aromatic. It was once used as a vegetable, and often indicates the past existence of a kitchen garden. Its main culinary use has now been taken over by celery. Note that occasionally there may be two small bracts.

b

a

Wild Parsnip *Pastinaca sativa* has many loose umbels of small yellow flowers. The oval winged seeds (**a** immature, **b** mature) are light and are dispersed by the wind. The root is not good to eat but it is the original species from which the domestic Parsnip was developed. It grows best in limy grass and wasteland.

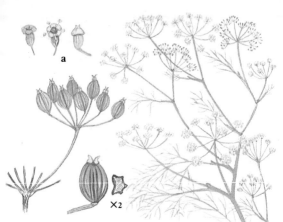

Fennel *Foeniculum vulgare* is noticeable with its feathery leaves in the coastal areas in which it has naturalised. The small umbels of yellow flowers with no bracts or bracteoles below the fruit will confirm identification. Three stages of development from bud to fruit are shown (**a**). If chopped finely the leaves are a good addition to salads, imparting a slightly aniseed flavour. If cooked with oily fish like mackerel or bass the oiliness is not so noticeable. Ice-cold soup made with yoghurt and fennel is delicious.

Rock Samphire *Crithmum maritimum* is another plant of the coast, sometimes literally on the beach. As with all such plants it is fleshy with narrowly lobed leaves. The small umbels of flowers and seeds have both bracts and bracteoles. It used to be pickled in brine for transporting to cities as an out of season green food, but not nowadays. The name derives from the patron saint of French fishermen, St. Pierre, which became anglicised as 'sampier' and then 'samphire'.

×2

Bilberry *Vaccinium myrtillus* is short, erect and deciduous. It is an undershrub of acid moorland. Note the angled green stem and finely toothed leaves. The flowers are pinkish with a green tinge; the berries are edible and best when cooked in a pie or stewed – they are rather acid when eaten raw. It is a fiddly job picking a quantity, but amply rewarding with the results.

Cowberry *Vaccinium vitis-idaea* also grows on moorland and has clusters of flowers. Note the leathery leaves with edges rolled under. It is prostrate and rooting along the stem. Sometimes it hybridises with Bilberry. The fruit is very acid and makes a good jelly, but you will need to add apple in order to introduce sufficient pectin to make it set.

Northern Bilberry *Vaccinium uliginosum*, as
its name implies, grows only in northern areas
of Britain – certainly not south of Yorkshire.
It is a smaller fruit than the previous species.
The twigs are round and brown as opposed
to angled and green, and the leaves are untoothed and blue-green. It is edible and di-
gestible in small quantities.

Cranberry *V. oxycoccos* is a creeping under-shrub. The leaves are dark green, oval and whitish on the underside. Note the prominent stamens on the flowers. The fruit is the rather acid berry making a delicious jelly to eat with poultry. It grows in boggy land but is now uncommon in Britain due to drainage of wetland. In the 16th century it was widely used in sauces. It is not edible when raw, being so acid. In Scandinavia this is put to good use in the cleaning of silver.

Alpine Bearberry *Arctostaphylos uva-ursi* (**1**) is prostrate and mat-forming. The leathery leaves are evergreen, and note that they are widest near the tip. It grows on high moorland, and when in woods will grow up to about 2 m. The berries are dry and floury and not very good to eat. They have been used medicinally in the past. The leaves are sometimes added to Coltsfoot when making a herbal smoking mixture. **Black Bearberry** *A. alpina* (**2**) is deciduous, and note the toothed leaves. It grows in high moorland and the berries are sweet and cloying.

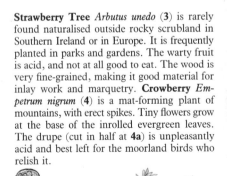

Strawberry Tree *Arbutus unedo* (**3**) is rarely found naturalised outside rocky scrubland in Southern Ireland or in Europe. It is frequently planted in parks and gardens. The warty fruit is acid, and not at all good to eat. The wood is very fine-grained, making it good material for inlay work and marquetry. **Crowberry** *Empetrum nigrum* (**4**) is a mat-forming plant of mountains, with erect spikes. Tiny flowers grow at the base of the inrolled evergreen leaves. The drupe (cut in half at **4a**) is unpleasantly acid and best left for the moorland birds who relish it.

177

Primrose *Primula vulgaris* (**1**) is the well-known plant of early springtime. It is rapidly becoming less common due to the depredations of those who illegally uproot it. The fruit is the base of the calyx tube (partly cut away at **1a**). **Cowslip** *P. veris* (**2**) fruit is enclosed within a much shorter and more enlarged calyx. Limestone grassy places favour it, but fertilisers and uprooting have removed it from many areas. **Oxlip** *P. elatior* (**3**) fruit is oblong, protruding from the calyx (**3a**). It is rare but locally abundant in some clayey woodlands. Note the following leaf shapes: (**1**) tapers to a very short petiole; (**2**) narrows abruptly to a winged petiole, about 15 cm long, and (**3**) is similar to (**2**) but longer – up to 20 cm long.

Water Violet *Hottonia palustris* (**4**) is not a violet but a primula. The leaves are feathery and submerged in the water in which it grows. The capsule (**4a**) splits open at the top. Drainage of wetlands and 'canalising' of ditches is making this delightful plant scarce. **Scarlet Pimpernel** *Anagallis arvensis* (**5**) is a low creeping plant of disturbed ground. The flowers open only in sunlight. Note the open capsule with the persistent style (**5a**). **Yellow Pimpernel** *Lysimachia nemorum* (**6**) is prostrate in shady woods. Note the 5 divisions in the capsule (**6a**), common to all three plants.

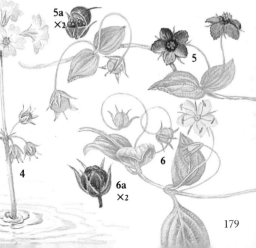

4a
×2

5a
×2

5

4

6

6a
×2

Sowbread *Cyclamen hederifolium* is more familiar as a garden plant or naturalised in parks. This species is rare in the wild, where it grows in woodland. It was introduced into Britain many centuries ago and its name derives from the relish with which wild boar rooted up the large round corms. The seed capsule opens by valves which fold back with the spirally coiled stalk.

Thrift *Armeria maritima* is a common plant of saltmarshes, sea cliffs and mountains. It grows with a long penetrating root stock and is adapted to habitats hostile to many plants. The papery calyx cups the fruit. The flowers dry well and are everlasting, making them subjects for flower arrangers.

a

(P) **Privet** *Ligustrum vulgare* is a shrub growing to 5 m. It is more or less deciduous. The leaves contain an irritant which causes inflammation in some delicate skins, and the berries are poisonous. These are two seeded (**a**) and have an oily flesh. The white flowers are fragrant and attract many insects. It grows best on calcareous soils in scrub and woods. It is not the species most commonly used as a hedgerow plant: this is *L. ovalifolium* (not illustrated), which has more persistent oval leaves. The leaves of *L. vulgare* produce a yellow dye and the berries a blue-green one.

Bogbean *Menyanthes trifoliata* grows in freshwater bogs everywhere. It is one of the most ancient of plants, having thrived before the Ice Age. The flowers have fringes of cotton-like hairs. The trifoliate leaves are like those of the broad bean. Creeping underwater stems enable it to colonise large areas; in garden ponds it is invasive but controllable. Note the persistent style on the fruit.

×½

Wild Madder *Rubia peregrina* (**1**) is a scrambling plant with rough angled stems. Down-turned prickles line the angles; the leaves are leathery and prickly. Unlike other plants in this family it has berries each containing one seed. It usually grows in scrubland and in coastal rocky places.

Common Cleavers *Galium aparine* (**2**), sometimes known as goosegrass, scrambles through hedgerow plants using the backward facing hooks on the square stem. To aid seed dispersal the fruit (**2a**) is similarly equipped. Said to be a good food for goslings and an indicator of fertile soil.

Houndstongue *Cynoglossum officinale* (**3**) is tall and grows in dry grassland. Note the four nutlets joined at the centre within the calyx; all are covered with bristles (**3a**). The plant smells of mice, and its name is derived from its mediæval reputation of curing bites of mad dogs. **Blue Gromwell** *Lithospermum purpurocaeruleum* (**4**) is rare in Britain. It grows in limy scrub or woods. Note the flower opens purple and turns blue as it matures. The fruit is a pair of nutlets which are white and shiny (**4a**).

3

4a

4

3a

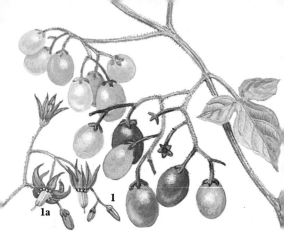

ℙ Bittersweet (Woody Nightshade) *Solanum dulcamara* (**1**) is the first of this family of plants with excellent vegetable members – such as potato and tomato – and also poisonous plants such as this. It grows in woods and in wasteground, sometimes scrambling in hedges. The reflexed petals and column of yellow anthers are similar in both these nightshade plants (**1a** and **2a**). Berries formed into garlands were said by Culpeper to be a remedy for dizziness. Legend credits them with power to ward off evil spirits, and this is borne out by the finding of many of them in Tutankhamen's third coffin. **ℙ Black Nightshade** *S. nigrum* (**2**) differs in having toothed leaves and a dull black, equally poisonous berry. It often appears as a weed of cultivation.

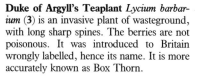

Duke of Argyll's Teaplant *Lycium barbarium* (**3**) is an invasive plant of wasteground, with long sharp spines. The berries are not poisonous. It was introduced to Britain wrongly labelled, hence its name. It is more accurately known as Box Thorn.

ⓅDeadly Nightshade *Atropa bella-donna* is another example of a naturally poisonous plant being used medicinally even today. It is a stout, well-branched, dull plant and grows to 150 cm. Note that the sepals are retained by the shiny black berry. You will find it in limy woods and scrub. One poison it contains is atropine, used in ophthalmic surgery for dilating the pupil of the eye. In the Middle Ages the ladies of Venice used it to emphasise their lovely dark brown eyes, making them 'bella donna' – beautiful ladies. It is used cosmetically in the same way today and causes many problems. Gerard named it a 'naughty and deadly plant'. It is not very common.

(P) Henbane *Hyoscyamus niger* is yet another very poisonous plant. It occurs throughout Britain on roadsides and cultivated or disturbed ground. The fruit (**a**) is enclosed within the swollen calyx (**b**) and releases seed through an opening after the cap is removed (**c**). The main poison is hyoscine, used in minute quantities to control suffering, but it causes delirium, spasms and death following large doses, and the whole plant must be treated with great care.

℗Thorn Apple *Datura stramonium* is not only poisonous but it is a dangerous narcotic. It grows to about 1 m, producing its spiny fruit. The whole plant has an unpleasant fœtid smell. The fruit splits open by four valves to release the seed. It grows on bare wasteground and is best left alone. Among the poisonous alkaloids it contains are hyoscine and scopolamine.

190

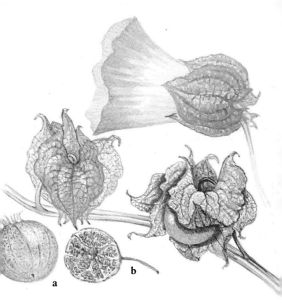

a

b

(P) The **Apple of Peru** *Nicandra physaloides* is a medium sized plant making a rounded shape, large for an
annual species. It is fœtid, hairless and poisonous.
Note a reddish tinge in the lower parts of the stem.
The sepals encase the brown berry when ripe (**a**).
The transverse section (**b**) shows the many small
seeds in the divided capsule.

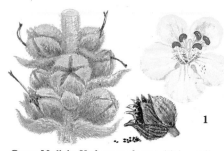

Great Mullein *Verbascum thapsus* (**1**) is a tall plant with grey-green, felted leaves. These will be seen early in the year before the spike arises, which will grow to 2 m. Note the way in which the flowers open at random on the spike. You will find it on sunny hedgebanks and open places. The capsule is slightly longer than the calyx, and turns brown when ripe to release the many small blackish-brown seeds.
Dark Mullein *V. nigrum* (**2**) is a smaller plant, sometimes branching. The oval seed capsule is much longer than the calyx. **White Mullein** *V. lychnitis* (**3**) has white flowers and a more elongated capsule.

×⅛

Common Figwort *Scrophularia nodosa* (**4**) is a fairly tall plant growing in damp hedgerows. It is more or less hairy towards the top of the plant only, and has a definite square stem. The flowers of these three figworts are similar. Note how the seed capsule opens. **Green Figwort** *S. umbrosa* (**5**) is quite hairless and seems winged because the leaf stalks run down the stem. **Balm-leaved Figwort** *S. scorodonia* (**6**) is downy all over and has an unwinged squarish stem. To confirm identification note the three different leaves.

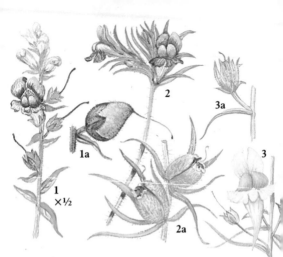

Snapdragon *Antirrhinum majus* (**1**) is best known as a garden plant of varying colours. This species was introduced many centuries ago as a wall plant and in the 18th century was cultivated for the garden, with many hybrids arising. Note the retained style and the capsule longer than the calyx (**1a**). **Lesser Snapdragon** *Misopates orontium* (**2**) is smaller, with linear leaves. The capsule is shorter than the calyx (**2a**). This is a weed of cultivation but the **Common Toadflax** *Linaria vulgaris* (**3**) is more often found in hedgerows and waste ground. The capsule is more than twice as long as the calyx (**3a**) and the seed is winged and is scattered by the wind.

Foxglove *Digitalis purpurea* (**4**) is a plant containing a drug that has perhaps saved more lives than most others. It quickly colonises wasteland with its familiar tall spikes. The seed capsules, turning brown on the stem, release the very fine seed each time the plant moves. In the late 18th century a doctor botanist, William Withering, noted the value of digitalin contained in the plant in the treatment of heart diseases. It is still an accepted and valuable drug within the British Pharmacopoea and certain species of Foxglove are cultivated for the purpose of its extraction.

4

The **Monkey Flower** *Mimulus guttatus* (**1**) is a
water's edge plant introduced to British gardens
in the 19th century, and now widely naturalised
along watercourses. Two smaller species, *M.
lutea* and *M. moschatus*, occur less frequently.
The fruit (**1a**) contained in the inflated sepal
tube (**1b**) opens down the middle of each cell of
the ovary. **Yellow Rattle** *Rhinanthus minor* (**2**) is
semi-parasitic on grasses. The winged seed (**2a**)
is loose within the inflated calyx (**2b**), giving the
plant its various local names. They include Baby
Rattle and Pennygrass. Gerard said it was of 'no
account and useless'. It makes a very pleasant
addition to some waste grassland.

3a
×2

Plantains grow from a central rosette of ribbed leaves. The flowers open progressively from the bottom of the spike with prominent anthers; fruit is a capsule with a 'lid' opening to release the seed (**3a**). The leaves of **Greater Plantain** *Plantago major* (**3**) abruptly narrow to a longish petiole; anthers start lilac, becoming yellow. **Hoary Plantain** *P. media* (**4**), the only scented one, has leaves gradually narrowing to a short petiole. Anthers are pink. **Ribwort** *P. lanceolata* (**5**) has lanceolate leaves with a short petiole, and yellow anthers. **Buckshorn Plantain** *P. coronopus* (**6**) has pinnate leaves, often red-tinged, and yellow-green anthers. **Sea Plantain** *P. maritima* (**7**) has narrow fleshy leaves showing 3-5 veins faintly, and yellow or pinkish flowers.

3

6

4

7

5

×⅙   ×⅙   197

1   **1a**   **1b**

Elder *Sambucus nigra* (**1**) is a shrub found in waste-land and roadsides throughout Britain and Europe. It grows well branched to 10 m (**1a**). The lower wood is hard, whereas that in the upper branches is soft with a very spongy pith. The leaves are pinnate with toothed leaflets. It has a smell, some say, of cats. A spray of leaves carried is said to be a good fly deterrent. The flowers are fragrant and grow in flat-topped clusters and are pollinated by flies. They make excellent white wines, which may be still or sparkling, and used to flavour a sorbet they are delicious. As recently as 1950 they were still listed in the British Pharmacopoeia as the source of a good eye lotion. The tiny gooseberry-like fruit (**1b**) develops into racemes of shiny black berries on red stalks. As they ripen so they turn and hang down. For centuries they have been used for making most delicious red wines. **Red-berried Elder** *S. racemosa* (**2**) is often planted and has sometimes become naturalised in hilly areas. The flowers are yellow-green and the leaflets narrow and toothed. Note the more rounded cluster of berries, on stalks not as red as (1). **Danewort** or **Dwarf Elder** *S. ebulus* (**3**) is a herb rather than a shrub, growing to about 1 m. It is fœtid and forms patches in scrub and roadside habitats. The flowers are pink-tinged.

(P) Guelder Rose *Viburnum opulus* grows in damp places. The flowers are in a flat cluster with larger flowers around the edge than in the main area; the large ones are sterile. It grows to 4 m. Note the few small round glands on the petiole at the base of the leaf blade. The leaves colour well in the autumn, so many have been planted in parks and gardens. The whole plant – berries, leaves and bark – is poisonous. Note the retained 'stump' of the style which is absent in red currant: the two must not be mistaken. A garden variety (var. *roseum*), the Snowball Tree, originally from Holland, has all the sterile flowers in a globose cluster.

Wayfaring Tree *Viburnum lantana* grows to about the same size but the flowers are in more rounded tight clusters and are all of the same size. It grows in scrubland and hedgerows on lime. The wrinkled leaves are oval and fine toothed. The berries start green and ripen through red to black. They are very astringent and are best left to birds, who will take them when other more palatable ones are finished. It is said the berries have been used to make a hair dye as well as ink. It is not found naturalised north of Yorkshire.

ⓅHoneysuckle *Lonicera periclymenum* (**1**) twines clockwise up trees and shrubs, often distorting saplings. The leaves are often the first to appear, being open during December for the following year. The red berries are poisonous. **Fly Honeysuckle** *L. xylosteum* (**2**) is a bushy shrub with smaller flowers which grow in pairs. They are not scented, and arise from the leaf axils near the ends of the stems. It is rare in Britain but you will find it in woodland in the south-east.

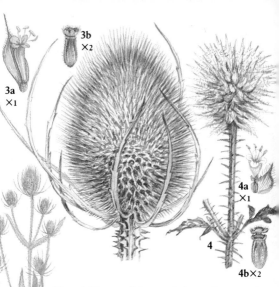

Teasel *Dipsacus fullonum* (**3**) is a tall plant of verges and damp ditches in southern Britain. It is prickly with angled stems. The little mauve flowers (**3a**) have long bracteoles which turn brown and brittle as the fruit (**3b**) ripens. The heads used by weavers to bring up the nap of cloth come from Fuller's Teasel *D. fullonum* ssp. *sativus* (not illustrated). The **Small Teasel** *D. pilosus* (**4**) is rare in similar habitats. Note the hairy bracteoles (**4a**) and the fruit (**4b**) with calyx 'lid'.

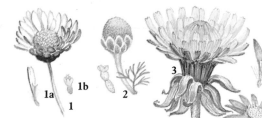

Introduction to the Daisy family

There are three basic flower forms in the Daisy family. For example the Daisy (**1**) has ray florets (**1a**) on the perimeter and disc florets (**1b**) in the centre. Pineapple Weed (**2**) has only disc, and Dandelion (**3**) only ray florets. To confuse us Sea Aster (**4**) has both, but often only disc-type (**4a**). The fruit is an achene, usually attached to a parachute of fine hairs called the pappus. The daisy achene (**5**) is very small and hairy but without a pappus. The hairs of the pappus may be feathery (shown below, many times magnified) like those of the Hawkbits (**6**) or simple like those of the Rough Hawksbeard (**7**), which is joined directly to the achene – it has no beak. Beaked Hawksbeard has a short beak (**8**), Stinking Hawksbeard (**9**) a long one.

Bracts on fruiting heads of some species are helpful in identifying them. Burdock (**10**) has hooked bracts, but note those of Black Knapweed (**11**), Brown Knapweed (**12**) and the purplish tipped bracts of Cornflower (**13**) and Sawwort (**14**). Thistles have spiny-tipped bracts, small on Slender Thistle (**15**) or long and recurved on Musk Thistle (**16**). Check the size and shape of the pappus which varies from small, white and in a clock on dandelion (**17**) to a fawn untidy mass on Creeping Thistle (**18**), which is a fragrant plant, attractive to many butterflies (**19**). Thistledown in the wind from Spear Thistle (**20**) is a familiar sight, as is the goldfinch feeding on thistle seeds.

Coltsfoot *Tussilago farfara* (**1**) is one of the very early plants flowering in spring. You will find it growing on wasteground and sometimes as a dreadful weed of cultivation. Note that the erect head droops after pollination and then re-erects on a lengthening stem as the seed ripens. It has both disc (**1a**) and ray (**1b**) florets which close up at night. The white pappus which forms a 'clock' is composed of simple hairs and is unbeaked. Goldfinches find it a good supply of early food. The leaves appear after the flowering stalk has died down.

Butterbur *Petasites hybridus* (**2**) forms patches in wet grassland by means of offshoots. Male (**2a**) and female flowers grow on separate plants. The stem of the fruiting head elongates as it ripens and the unbeaked white pappus of long simple hairs opens (**2b**). After the flowers have died the leaves, which may be up to 1 m across, appear. The leaves were once used to wrap dairy produce – cheese and butter – hence the name. They have also been used as sun and rain hats. **Winter Heliotrope** *P. fragrans* (**3**) is often found by roadsides and on wasteland by following the scent. The leaves appear in November and are soon followed by the flowers. The pappus and the elongating fruiting stalk (**3a**) are similar to (**2**).

Groundsel *Senecio vulgaris* (**1**) is to be found in any cultivated ground not sterilised by chemicals. It is a weak-growing plant which spreads rapidly through gardens, but being shallow-rooted it is easily removed. It rarely has ray florets. The pappus is white with long simple hairs and attached directly to the achene.

Carline Thistle *Carlina vulgaris* (**2**) grows on limestone scrub and grassland. It is a short erect plant with spiny leaves. The florets are all tubular and form a flattish cone. Inner straw-coloured bracts which look like ray florets spread open in dry weather. They survive the winter and make the plant a good subject for dried decorations. The pappus is of feathery hairs. Its habit of opening and closing in dry and wet weather has earned it a place as a country barometer.

Lesser Burdock *Arctium minus* (**3**) is large with rough purple-veined leaves, and grows in hedgerows and

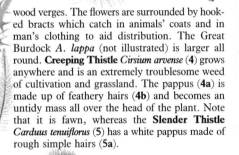

wood verges. The flowers are surrounded by hooked bracts which catch in animals' coats and in man's clothing to aid distribution. The Great Burdock *A. lappa* (not illustrated) is larger all round. **Creeping Thistle** *Cirsium arvense* (**4**) grows anywhere and is an extremely troublesome weed of cultivation and grassland. The pappus (**4a**) is made up of feathery hairs (**4b**) and becomes an untidy mass all over the head of the plant. Note that it is fawn, whereas the **Slender Thistle** *Carduus tenuiflorus* (**5**) has a white pappus made of rough simple hairs (**5a**).

1a

1b

Goatsbeard *Tragopogon pratensis* (**1**) has grass-like leaves so is often overlooked when not flowering or in seed. Note the long bracts – longer than the ray florets when the flower opens in early morning. It closes again around midday. The achenes have a long beak below the very feathery hairs (**1a**). The pappus makes a very large light clock (**1b**). It is quite common on road verges of south-east England and throughout Europe. It was known by the ancient Greeks as *barba hirci*, which William Turner (sometimes called the 'father of botany') translated literally as goat's beard.

1
×⅛

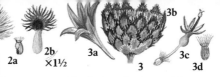

It was suggested in the introduction to this section (page 205) that the bracts of some species be noted. **Black Knapweed** *Centaurea nigra* (**2**), often called Hardhead, is common. Note the very short pappus hairs on the achene (**2a**) and the bracts (**2b**). **Cornflower** *C. cyanus* (**3**) is rarely found in Britain except in gardens. The purple-tinted involucral bracts (**3b**) support the ray florets (**3a**) and central ones (**3c**). A short pappus only tops the achene (**3d**). **Greater Knapweed** *C. scabiosa* (**4**) is the largest of the genus. There are infertile ray florets and the viable central ones (**4a**) which have a small pappus. The bracts are green (**4b**). When all the achenes have left the receptacle it shows a bright golden colour (**4c**) and dries well for winter arrangements. **Red Star Thistle** *C. calcitrapa* (**5**) is easily recognised by its long pointed bracts. It has no pappus.

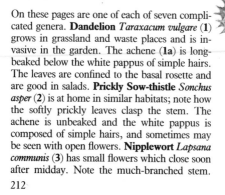

1

1a
×2

On these pages are one of each of seven compli-
cated genera. **Dandelion** *Taraxacum vulgare* (**1**)
grows in grassland and waste places and is invasive in the garden. The achene (**1a**) is long-
beaked below the white pappus of simple hairs.
The leaves are confined to the basal rosette and
are good in salads. **Prickly Sow-thistle** *Sonchus
asper* (**2**) is at home in similar habitats; note how
the softly prickly leaves clasp the stem. The
achene is unbeaked and the white pappus is
composed of simple hairs, and sometimes may
be seen with open flowers. **Nipplewort** *Lapsana
communis* (**3**) has small flowers which close soon
after midday. Note the much-branched stem.

The achenes are tightly packed into the involucre with no pappus (**3a**); it dries well (**3b**). **Prickly Lettuce** *Lactuca serriola* (**4**) has violet-tipped bracts (**4a**) which open to release the beaked achene with its pappus of simple white hairs (**4b**). **Wall Lettuce** *Mycelis muralis* (**5**) is similar but the achene is abruptly beaked and the pappus has short hairs on the edge. **Autumn Hawkbit** *Leontodon autumnalis* (**6**) is found on verges and wasteland. Note the leaves and black hairs on the bracts. The pappus is of feathery hairs (**6a**) sometimes with a row of simple hairs surrounding them. The achene is unbeaked. **Common Catsear** *Hypochaeris radicata* (**7**) grows in dry grassy places. Note the hairy leaves in the flat rosette and the small scale-like leaves on the stem. The pappus is tightly enclosed before being released. The achene has a short beak and the pappus feathery hairs (**7a**).

Three more groups are represented here by one each. **Leafy Hawkweed** *Hieracium umbellatum* (**1**) is tall, unbranched and covered with soft hairs. The achene is unbeaked below a fawn-coloured pappus of simple hairs. The bracts on the involucre are blackish-green. **Hawkweed Ox-tongue** *Picris hieracioides* (**2**) is closely related. Note that it is bristly all over. It has a white pappus of feathery hairs over beaked achenes. It grows on field edges, verges and waste on calcareous soil. **Smooth Hawksbeard** *Crepis capillaris* (**3**) was introduced to Britain in uncleaned seed corn. It grows mainly on wasteground, and is branched with shiny leaves. The pappus is white and does not form a clock. Note the curved achene.

ⓅBlack Bryony *Tamus communis* (**4**) twines clockwise up hedge plants to about 4 m. The berries and the black fleshy root are very poisonous. Female flowers (**4a**) and male ones grow on separate plants. The berries persist through the winter. Red berries at the foot of the hedge or in woodland boundaries will be **ⓅLords and Ladies** *Arum maculatum* (**5**). The leaves appear in December before spring brings forth the unusual flower with the purple spadix. The column of red berries, ripening from green, starting at the top, is poisonous. A similar plant, Large Cuckoopint *A. italicum* (not illustrated) has a yellow spadix.

4a

4

5

℗ **Lily of the Valley** *Convallaria majalis* (**1**) grows in dry woodland in limy areas. The fragrant white flowers (**1a**) appear in spring and the berries, which are poisonous, ripen in autumn – but this is uncommon in Britain. **Bog Asphodel** *Narthecium ossifragum* (**2**) grows in acid wetland in spring, the flowers adding a splash of golden colour (**2a**) to moorland bogs. In autumn the orange-coloured capsule and stems take their place. **Fritillary** *Fritillaria meleagris* (**3**) naturalises well in garden grassy places. It may be dull purple or whitish. The capsule splits into three parts (**3a**) to release the seed. Agricultural chemicals and drainage destroy many plants each year.

Many woodland floors in spring are a blue haze of **Bluebell** *Hyacinthoides non-scripta* (**4**). The flowers (**4a**) with two bracts give way to the swollen ovary (**4b**) which is retained into the summer to ripen (**4c**). The section (**4d**) shows the seeds in the compartments. Plants are severely damaged when trodden on, so tread carefully when picking flowers. **Spring Squill** *Scilla verna* (**5**) is a little plant with single bracts and leaves all from the roots. Capsules are three-sided, more or less globular. **Autumn Squill** *S. autumnalis* (**6**) flowers before the leaves appear and has no bracts. The capsule is more oval. Both squills are plants of coastal and mountainous turf.

Meadow Saffron *Colchicum autumnale* (**1**) blooms in the autumn before the leaves appear. The lower parts of the petals are fused together, resembling a stem. As the fruit capsule ripens the leaves grow in large clumps in early spring. You will find it in damp meadows and light woodland. It is not the plant from which saffron is obtained – this is the Saffron Crocus *Crocus sativa* (not illustrated). **Herb Paris** *Paris quadrifolia* (**2**) also grows in woodland habitats, but only on calcareous soils. The four leaves (**2a**) are easily recognised. The berry-like capsule with its long sepals splits to release its seeds. It is now rare and found only in eastern Britain.

2

2a
×⅕

1

3

Butcher's broom *Ruscus aculeatus* (**3**) is a much-branched bush with leaves sharply spined at the tips. The flower grows on the upper surface of the leaf, which is botanically called a cladode, or modified stem. Young shoots are said to taste like asparagus if cooked in the same way.

4

Wild Asparagus *Asparagus officinalis* (**4**) is rarely naturalised in coastal grassland. The tiny whitish flowers are easily missed but the red berries are conspicuous. It is widely cultivated for its decorative and vegetable value.

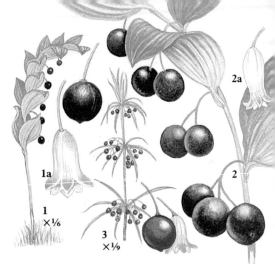

Solomon's Seals are most graceful plants and are often grown in gardens. **Angular Solomon's Seal** *Polygonatum odoratum* (**1**) has an angled stem and fragrant flowers (**1a**) growing singly and not waisted. **Common Solomon's Seal** *P. multiflorum* (**2**) is similar but has a round stem and waisted scentless flowers (**2a**) in groups of 2-5. The berries are similar in both species. **Whorled Solomon's Seal** *P. verticillatum* (**3**) is quite different in its leaf formation and red berries. It is very rare and specially protected under the Wildlife and Countryside Act. All are woodland plants. The berries are said to be non-poisonous and most unpleasant to eat.

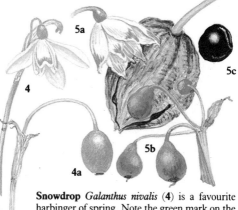

Snowdrop *Galanthus nivalis* (**4**) is a favourite
harbinger of spring. Note the green mark on the
inner petals only. The oval capsule (**4a**) droops
into the grass in which the plant grows and may
well be missed unless carefully searched for.
Summer Snowflake *Leucojum aestivum* (**5**) is
much rounder (**5a**). Its maturing capsules are
like little pears (**5b**) ripening to brown contain-
ers with large buoyant seeds (**5c**). These plants
are native along the banks of the Thames and
some other southern British rivers. The seed is
dispersed downstream by the flow of the water.

5
×⅓

Yellow Iris *Iris pseudacorus* grows in marshes, by and sometimes in water. The large, roughly ribbed capsule is evident soon after the petals have fallen in summer. It ripens in autumn and opens in three sections as the brownish seeds are ready to be released. It has rhizome root stock and spreads by this means, forming large and invasive patches.

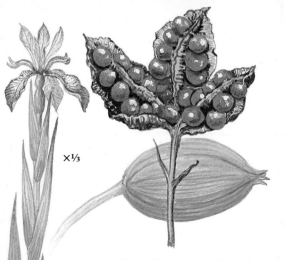

×⅓

Stinking Iris *I. foetidissima* is found in hedge-rows, scrub and woodlands, usually on limy soil. It gets its name from the sickly-sweet smell of its leaves when crushed. The scent also gives it its alternative name of Roast Beef Plant. The pod appears in late summer – note that it is smoother than the previous species – and opens in three sections as the bright orange seeds ripen. It is a good garden plant.

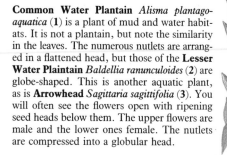

Common Water Plantain *Alisma plantago-aquatica* (**1**) is a plant of mud and water habitats. It is not a plantain, but note the similarity in the leaves. The numerous nutlets are arranged in a flattened head, but those of the **Lesser Water Plaintain** *Baldellia ranunculoides* (**2**) are globe-shaped. This is another aquatic plant, as is **Arrowhead** *Sagittaria sagittifolia* (**3**). You will often see the flowers open with ripening seed heads below them. The upper flowers are male and the lower ones female. The nutlets are compressed into a globular head.

Flowering Rush *Butomus umbellatus* (**4**) is a tall water plant growing erect from the bottom of ponds and river banks. Note that the leaves are three-cornered. The six carpels in the fruit (**4a**) are joined at the base. It is getting rare as rivers and streams are canalised.

Great Reedmace *Typha latifolia* (**5**) grows in swampy ground. The flat leaves overtop the flower heads. Male flowers (**5a**) are immediately above the females (**5b**), becoming a brittle spike following pollination. The seeds are small, dry and surrounded by a parachute of fine hairs (**5c**).

5
×1/15

5a

5b

5c

4
×1/10

4a

Branched Bur-reed *Sparganium erectum* (**1**) grows in slow-moving water and on river banks, forming large clumps which provide good cover and nest sites for many water birds. The individual flowers of either sex are insignificant and in globular heads. The many males above the few females on each plant show only yellowish stamens. When the pollen, which is windborne, is exhausted the male fruit head wilts and falls, exposing the angular branches. The female head then ripens with many-beaked dry seeds, giving them a burr look. The firm upright leaves are triangular in shape.

1

There is a wealth of beauty and fascinating information about European terrestial orchids and books have been written about them. Suffice it to say here that orchids rely on certain organisms in the soil which are not generally distributed. The chances that they will survive transplanting, even with the owner's permission, are minimal. Some orchids take many years to flower from seeds and only do so once before they die – they are monocarpic. As far as anything is typical in orchids it is the seed capsule (**2**) which opens by three valves (**3**) to release the dust-like seeds. The capsule of the **Marsh Helleborine** *Epipactis palustris* (**4**) is similar in form but droops when ripe. **Lesser Twayblade** *Listera cordata* (**5**) has many small flowers (**5a**) and globular red-ribbed capsules (**5b**). The twin leaves at the base have usually died off by this time but may still be visible. **Pyramidal Orchids** *Anacamptis pyramidalis* (**6**) are variable and are found in 'unimproved' grassland in the south. The many capsules release innumerable seeds through the valves.

These five plants are exotic introductions to Britain and found only rarely in a naturalised habitat. **Mulberry** *Morus nigra* (**1**) was brought from the east in an effort to build up a silk industry, but it turned out to be the wrong species. Silk worms feed on the White Mulberry *M. alba*. *The fruit is good to eat in early autumn.* **Snowberry** *Symphoricarpos rivularis* (**2**) is spreading rapidly by means of suckers. The berry is not good to eat and even birds leave it alone. It contains air and if trodden on is inclined to explode with a loud pop.

Firethorn *Pyracantha coccinea* (**3**) makes an excellent garden boundary plant and grows equally well up a wall. Being much favoured by birds the seeds from the red (another variety has orange) berries are sometimes dispersed into hedgerows. **Medlar** *Mespilus germanica* (**4**) has naturalised in a few hedges in the south from plants grown for their fruit. Note the prominent calyx. **Honesty** *Lunaria annua* (**5**) is a garden escaped species of crucifer. Note that the silicula are almost round, and are much favoured by flower arrangers for winter decorations.

Index of scientific names

231

233

Index of English names